The Essence Of Theorophy

The Essence Of Theorophy

Maria .M

Spectra Enterprise

Contents

INDEX   1

Chapter 1   3

Chapter 2   20

Chapter 3   36

Chapter 4   54

Chapter 5   69

Chapter 6   85

Chapter 7   102

Chapter 8   119

Chapter 9   136

INDEX

Chapter 1: Introduction

1.1 Brief overview of Theorophy and its roots in ancient philosophical and spiritual traditions.

1.2 Definition and exploration of the term "Theorophy."

1.3 Setting the stage for a journey into the essence of Theorophy.

Chapter 2: Origins and Historical Context

2.1 Tracing the historical development of Theorophy through various cultures and civilizations.

2.2 Examining key figures and texts that have contributed to the evolution of Theorophy.

2.3 Highlighting the cross-cultural influences that have shaped Theorophical thought.

Chapter 3: Core Principles of Theorophy

3.1 Identifying and explicating the fundamental principles that underlie Theorophy.

3.2 Exploring the interconnectedness of spirituality, philosophy, and mysticism within Theorophical teachings.

3.3 Illustrating how these principles provide a holistic understanding of existence.

Chapter 4: The Inner Journey

4.1 Delving into the concept of the inner journey in Theorophy.

4.2 Discussing meditation, contemplation, and self-discovery as essential practices.

4.3 Exploring the transformative nature of the inner journey and its impact on personal growth.

Chapter 5: Theorophical Ethics

5.1 Examining the ethical framework within Theorophy.

5.2 Discussing the principles of compassion, kindness, and interconnectedness.

5.3 Exploring how Theorophical ethics can guide individuals in their daily lives

.

Chapter 6: Unity of All Religions

6.1 Investigating the concept of the unity of all religions in Theorophy.

6.2 Highlighting common threads among various religious traditions.

6.3 Discussing how Theorophy promotes interfaith understanding and harmony.

Chapter 7: Theorophy and Science

7.1 Exploring the relationship between Theorophy and scientific inquiry.

7.2 Discussing how Theorophical principles align with certain scientific concepts.

7.3 Addressing the potential for a harmonious integration of spirituality and science.

Chapter 8: Mystical Experiences in Theorophy

8.1 Examining the role of mystical experiences in Theorophical practices.

8.2 Exploring the nature of spiritual revelations and insights.

8.3 Discussing how mysticism enhances the understanding of Theorophical principles.

Chapter 9: The Future of Theorophy

9.1 Speculating on the relevance of Theorophy in the modern world.

9.2 Discussing potential applications of Theorophical principles in addressing contemporary challenges.

9.3 Encouraging readers to explore Theorophy as a source of wisdom and guidance.

Chapter 1

Introduction

The embodiment of Theorophy lies at the crossing point of reasoning, religious philosophy, and supernatural quality, winding around an embroidery that looks to unwind the secrets of presence and the idea of reality itself. Established in the old insight of different profound customs, Theorophy rises above the limits of regular idea, welcoming searchers to investigate the profundities of awareness and the interconnectedness, everything being equal.

At its center, Theorophy is a comprehensive way to deal with understanding the basic inquiries that have bewildered humankind for quite a long time. It isn't bound to a particular strict regulation but instead draws motivation from the all inclusive strings that go through different conviction frameworks. Theorophy coaxes people to leave on an excursion of self-revelation, digging into the domains of both the seen and the inconspicuous.

One of the vital mainstays of Theorophy is the investigation of transcendentalism, the part of reasoning that dives into the idea of reality past the actual domain. Theorophy places that there is something else to presence besides what might be immediately obvious, and it urges specialists to scrutinize the actual texture of the real world. It digs into the idea of cognizance, setting that it isn't just a side-effect of the mind however a principal part of the actual universe.

In the journey for understanding, Theorophy acquires from the rich woven artwork of magical customs found across societies. Whether it be the Sufi spiritualists of Islam, the Kabbalists of Judaism, or the scrutinizing customs inside Christianity and Hinduism, Theorophy perceives the ongoing idea of magical experience that goes through these assorted ways. Magic, in the Theorophical sense, isn't restricted to a particular strict setting yet is viewed as an immediate experience with the otherworldly, a fellowship with the heavenly that rises above the constraints of language and doctrine.

Theorophy likewise draws in with the lasting way of thinking, an idea that proposes there is a center arrangement of general insights that underlie all strict and otherworldly practices. This viewpoint states that underneath the superficial variety of convictions and practices, there exists a consistent idea of intelligence that addresses the fundamental idea of the human experience and the hidden solidarity of all presence. Theorophy, in embracing the enduring way of thinking, looks to distil these general bits of insight and apply them to the contemporary difficulties of the cutting edge world.

Vital to Theorophy is the idea of internal change. It sets that genuine comprehension can't be accomplished exclusively through scholarly request however requires a significant change in cognizance. This change isn't just a singular pursuit yet has more extensive ramifications for society overall. Theorophy imagines a world where people, through their internal arousing, add to the aggregate development of humankind.

Theorophy doesn't avoid the intricacies of human life, recognizing the interaction of light and shadow inside the individual and society. It perceives the inborn duality of the human experience — the consistent dance between the egoic self and the higher self, among obliviousness and edification. Instead of denouncing the shadow perspectives, Theorophy considers them to be open doors for development and self-revelation, necessary to the excursion towards completeness.

In investigating the substance of Theorophy, one experiences the idea of the Outright — an unutterable, extraordinary reality that lies outside the ability to comprehend of human cognizance. Theorophy places that the Outright is the source and embodiment of all that is, the unmanifest ground from which the manifest world arises. It is the constant reality in the midst of the always changing peculiarities of the material world.

Theorophy underscores the experiential component of otherworldliness, empowering direct experiences with the heavenly instead of dependence on handed down information. It perceives that language and ideas, while significant apparatuses, are restricted in catching the totality of the extraordinary experience. Theorophy, accordingly, welcomes people to go past the limits of language and enter the domain of direct mysterious experience, where the limits between the knower and the known disintegrate.

The pith of Theorophy is entwined with the idea of solidarity cognizance, which declares that at a more profound degree of the real world, everything is interconnected. This point of view difficulties the dualistic perspective that sees partition among self and other, individual and aggregate, human and nature. Theorophy imagines a reality where people perceive their inborn interconnectedness with all of creation, encouraging a feeling of obligation and stewardship for the prosperity of the whole snare of life.

Theorophy draws in with the enduring inquiry of theodicy — the test of accommodating the presence of misery and evil with the idea of a big-hearted and transcendent heavenly. It perceives that human comprehension is restricted and that the secrets of presence may not be completely fathomable inside the bounds of the human psyche. Theorophy supports modesty even with the obscure, welcoming people to embrace the secret as opposed to capitulating to the arrogance of guaranteeing outright sureness.

In the investigation of Theorophy, the idea of time takes on an alternate aspect. It challenges the straight and restricted viewpoint of time, recommending that the past, present, and future are interconnected in an immortal dance. Theorophy places that the everlasting second is open to the individuals who rise above the impediments of customary discernment, giving a brief look into the immortal domain where all prospects coincide.

Theorophy perceives the extraordinary force of images and paradigms, understanding that they act as extensions between the limited and the endless, the cognizant and the oblivious. Images, whether tracked down in strict iconography, folklore, or dreams, go about as windows into the more profound layers of the mind and the aggregate oblivious. Theorophy empowers the consideration and understanding of images for of opening the secret insight encoded inside them.

At its heart, Theorophy is a way of affection and empathy. It perceives that a definitive the truth is certainly not a theoretical idea to be analyzed mentally however a living, throbbing power that penetrates all of creation. Love, in the Theorophical sense, isn't simply a feeling however the actual texture of presence — the binding together power that ties everything together. Theorophy calls people to typify love in their viewpoints, words, and activities, perceiving that genuine otherworldliness is indivisible from a caring approach to being on the planet.

The quintessence of Theorophy isn't restricted to the domain of exclusive or dynamic hypothesis; it has down to earth suggestions for day to day existence. Theorophy imagines a general public wherein the quest for material abundance is offset with a more profound comprehension of the otherworldly components of life. It requires an incorporation of science and otherworldliness, perceiving that both are fundamental parts of the human journey for information and understanding.

Theorophy recognizes the significance of variety in the embroidered artwork of human experience. It perceives that various people might reverberate with various ways and customs, and there is nobody size-fits-all way to deal with otherworldliness. The embodiment of Theorophy lies in embracing this variety while perceiving the fundamental solidarity that rises above social and strict limits.

In the excursion of Theorophy, the idea of edification takes on a nuanced and complex importance. It isn't just a scholarly seeing yet a lived insight of enlivening to one's real essence. Theorophy perceives that illumination is a continuous interaction, a non-stop unfurling of higher conditions of cognizance and a developing of shrewdness.

Theorophy empowers a reconsideration of the idea of self, rising above the restricted egoic personality and perceiving the more profound, all inclusive self that is interconnected with all of presence.

This change in self-discernment has significant ramifications for how people connect with themselves, others, and the world in general. Theorophy imagines a world wherein the egoic designs that sustain division and struggle are risen above, leading to a more agreeable and interconnected approach to being.

The embodiment of Theorophy lies in the joining of profound standards into the texture of day to day existence. It's anything but a way of thinking to be restricted to the ivory pinnacle of dynamic idea however a lived reality that pervades all parts of presence. Theorophy welcomes people to typify the immortal insight they look for, becoming vessels of affection, empathy, and astuteness on the planet.

In the investigation of Theorophy, the job of otherworldly practices becomes fundamental. Whether it be reflection, supplication, consideration, or custom, these practices act as vehicles for the immediate experience of the heavenly. Theorophy perceives that hypothesis alone is lacking; it should be supplemented by immediate, experiential commitment with the secrets of presence.

Theorophy isn't a creed to be forced yet a challenge to investigate the profundities of one's own being. It perceives that the excursion of self-revelation is extraordinary for every person, and there is nobody size-fits-all way to deal with the secrets of presence. Theorophy praises the variety of human experience and urges people to track down their own valid way.

In considering the substance of Theorophy, the idea of mystery holds a focal spot. It perceives that the idea of the truth is in many cases confusing and that apparently disconnected insights can exist together. Theorophy welcomes people to embrace the secrets and inconsistencies of presence, rising above the constraints of straight and twofold reasoning.

Theorophy isn't a call to pull out from the world however a challenge to completely draw in with it more. It perceives that otherworldliness isn't a departure from the difficulties of life yet a method for exploring them with shrewdness and effortlessness. Theorophy imagines a world where people, grounded in otherworldly standards, effectively add to the improvement of society and the mending of the planet.

In the substance of Theorophy, the idea of holy nature arises — an acknowledgment of the interconnectedness among otherworldliness and the climate. Theorophy supports a change in cognizance that recognizes the holiness of all of creation, cultivating a feeling of love and stewardship for the Earth. It imagines an agreeable connection among mankind and the normal world, perceiving that the prosperity of one is indivisible from the prosperity of the other.

Theorophy is definitely not a decent objective however a dynamic and developing excursion. It perceives that the comprehension of truth is a nonstop course of unfurling and that profound development is a long lasting pursuit.

Theorophy welcomes people to stay open to new experiences, to scrutinize their own suspicions, and to embrace the continuous experience of self-disclosure.

As we dig into the quintessence of Theorophy, we track down an encouragement to rise above the constraints of customary discernment and impression the endless conceivable outcomes that lie past. It is a call to investigate the secrets of presence with an open heart and an unassuming brain, perceiving that a definitive truth isn't an objective however an excursion — a steadily unfurling investigation of the vast profundities of cognizance. In the pith of Theorophy, we track down an encouragement to hit the dance floor with the heavenly, to partake in the continuous orchestra of creation, and to epitomize the immortal insight that has reverberated through the ages.

1.1 Brief overview of Theorophy and its roots in ancient philosophical and spiritual traditions.

Theorophy, a term got from the Greek words "theos" (god) and "philo" (love or insight), addresses an extensive way to deal with figuring out the essential idea of presence. This amalgamation of reasoning, philosophy, and mystery tries to disentangle the secrets of the real world and human awareness. Established in old philosophical and profound practices, Theorophy draws from a different exhibit of shrewdness found across societies and ages, giving a comprehensive system to investigating the profundities of presence.

To appreciate the pith of Theorophy, it is vital for follow its underlying foundations back to the rich embroidered artwork of antiquated philosophical practices. The Greek time frame, described by the prospering of Greek way of thinking, laid the foundation for some thoughts that later impacted Theorophical thought. The Dispassionate and Neoplatonic schools, for example, dug into supernatural requests about the idea of the real world, the presence of an extreme reality past the material world, and the connection between the physical and mystical domains.

Pythagoras, an old Greek logician and mathematician, likewise assumed a urgent part in molding Theorophical points of view. His investigation of the enchanted meaning of numbers and the idea of an extraordinary numerical request fundamental the universe reverberates with Theorophical topics of interconnectedness and general standards.

As Theorophy winds through time, it experiences the philosophical legacy of the East, especially inside Hinduism and Buddhism. Old Indian methods of reasoning, like Vedanta and Samkhya, consider the idea of extreme reality, oneself, and the interconnectedness, everything being equal. The Upanishads, essential messages in Vedanta, investigate the possibility of Brahman — a definitive, shapeless reality that underlies and binds together the variety of the manifest world.

Buddhism, with its accentuation on the idea of misery and the way to edification, adds to's how Theorophy might interpret the human condition. The investigation of awareness, fleetingness, and the interconnectedness of all peculiarities in Buddhist idea lines up with Theorophical points of view on the extraordinary idea of the real world.

Theorophy's foundations additionally reach out into the enchanted customs of the Center East. Sufism, the supernatural element of Islam, gives bits of knowledge into the immediate experience of the heavenly and the excursion of the spirit toward association with the Outright. The compositions of spiritualists like Rumi and Ibn Arabi reverberate with Theorophical topics, accentuating adoration, examination, and the groundbreaking force of profound practices.

Inside Judaism, the Kabbalistic practice offers an enchanted translation of the

Torah, investigating the idea of God, the production of the universe, and the way to otherworldly climb. The emblematic language of Kabbalah, with its accentuation on the Tree of Life and the interconnectedness of heavenly radiations, lines up with Theorophical ideas of holy imagery and all inclusive bits of insight.

Christian magic, from the compositions of the Desert Fathers to the archaic spiritualists like Meister Eckhart and Julian of Norwich, adds to the Theorophical embroidery. The accentuation on direct fellowship with the heavenly, the extraordinary force of adoration, and the association of the spirit with God reverberates with Theorophical subjects of inward arousing and the greatness of the egoic self.

In the East, Chinese and Daoist methods of reasoning additionally leave their engraving on Theorophical thought. Daoism, with its accentuation on the Dao (the Way) and the normal progression of presence, lines up with Theorophical thoughts of congruity, interconnectedness, and the quest for a healthy lifestyle.

Theorophy recognizes the lasting way of thinking — an idea proposing that there exists a center arrangement of general bits of insight fundamental all strict and profound practices. This point of view places that, past the variety of social articulations, there is a consistent idea of intelligence that addresses the fundamental idea of the human experience and the interconnected solidarity of all presence.

As Theorophy coordinates these different strands of old insight, it accentuates the experiential element of otherworldliness. Theorophical lessons empower direct experiences with the extraordinary instead of dependence exclusively on scholarly comprehension. The spiritualist's excursion, independent of social or strict setting, is viewed as an immediate fellowship with the heavenly — an extraordinary and binding together experience that rises above the limits of language and creed.

Key to Theorophy is the idea of transcendentalism — the part of reasoning worried about the idea of reality past the discernible, actual world. Theorophy places that the material domain isn't the sole reality; rather, it is an indication of more profound, otherworldly standards. The investigation of cognizance, the idea of oneself, and the connection between the unmanifest and manifest elements of reality comprises a center part of Theorophical request.

Theorophy draws in with the idea of the Outright — an unspeakable, otherworldly reality that fills in as the source and embodiment of all that exists. This Outright, whether alluded to as Brahman, the One, or by different names in different customs, is placed as the constant reality in the midst of the steadily changing peculiarities of the material world. Theorophy ponders the idea of this Extreme Reality and its suggestions for the comprehension of self and universe.

In the excursion of Theorophy, the idea of solidarity awareness arises. This point of view declares that, at a more profound degree of the real world, everything is interconnected. Theorophy challenges the dualistic perspective that sees partition among self and other, individual and aggregate, human and nature. The interconnectedness of all presence turns into a focal subject, encouraging a feeling of obligation and stewardship for the prosperity of the whole snare of life.

Theorophy likewise draws in with the idea of time, testing the customary direct viewpoint. It proposes that the past, present, and future are interconnected in an immortal dance. Theorophy places the presence of an everlasting second open to the people who rise above conventional insight, giving a brief look into an immortal domain where all prospects coincide.

Images and paradigms hold a huge spot in Theorophical thought. Perceiving the extraordinary force of images, whether tracked down in strict iconography, folklore, or dreams, Theorophy empowers the thought and translation of these images for of opening the secret insight encoded inside them. Images act as extensions between the limited and the boundless, the cognizant and the oblivious, offering windows into the more profound layers of the mind and the aggregate oblivious.

Theorophy recognizes the groundbreaking capability of internal work and self-disclosure. It perceives the interaction of light and shadow inside the individual and society. Instead of denouncing the shadow viewpoints, Theorophy considers them to be open doors for development and self-disclosure — basic to the excursion towards completeness. Theorophical lessons underscore the significance of coordinating and rising above the dualities intrinsic in the human experience, encouraging a fair and agreeable way to deal with life.

In investigating the substance of Theorophy, the idea of affection and sympathy arises as a focal topic. Love, in the Theorophical sense, isn't simply a feeling yet the actual texture of presence — the bringing together power that ties everything together.

Theorophy calls people to exemplify love in their viewpoints, words, and activities, perceiving that genuine otherworldliness is indistinguishable from an empathetic approach to being on the planet.

Theorophy is definitely not a separated, theoretical way of thinking yet a common-sense aide for carrying on with a significant and deliberate life. It imagines a general public where people, grounded in profound standards, effectively add to the prosperity of humankind and the planet. The quest for material abundance is offset with a more profound comprehension of the otherworldly elements of life, cultivating an agreeable joining of material and profound qualities.

Variety is a sign of Theorophical thought. Theorophy perceives that various people might resound with various ways and customs. It recognizes the variety of ways to deal with the heavenly and supports a conscious investigation of different otherworldly roads. The substance of Theorophy lies in embracing this variety while perceiving the hidden solidarity that rises above social and strict limits.

Theorophy draws in with the lasting inquiry of theodicy — the test of accommodating the presence of torment and evil with the idea of a big-hearted and transcendent heavenly. It recognizes the restrictions of human comprehension and energizes modesty despite the unexplored world. Theorophy welcomes people to embrace the secret instead of surrendering to the arrogance of asserting outright sureness.

In the substance of Theorophy, the idea of illumination takes on a nuanced and multi-layered significance. It isn't simply a scholarly seeing yet a lived insight of enlivening to one's real essence. Theorophy perceives that edification is a continuous interaction, a nonstop unfurling of higher conditions of cognizance and an extending of shrewdness.

Theorophy empowers a reconsideration of the idea of self, rising above the restricted egoic character and perceiving the more profound, all inclusive self that is interconnected with all of presence. This change in self-discernment has significant ramifications for how people connect with themselves, others, and the world in general. Theorophy imagines a world where the egoic designs that sustain division and struggle are risen above, leading to a more amicable and interconnected approach to being.

1.2 Definition and exploration of the term "Theorophy."

Theorophy, a term got from the Greek words "theos" (god) and "philo" (love or insight), addresses a complex and thorough way to deal with grasping the major idea of presence. At its center, Theorophy includes a union of reasoning, philosophy, and mystery, winding around together strings from different profound practices and old insight to make a comprehensive structure for investigating the secrets of the real world and human cognizance.

To characterize Theorophy is to set out on an excursion that rises above the limits of ordinary idea, welcoming people to investigate the profundities of presence with a receptive outlook and an open heart. The actual term proposes an affection or intelligence relating to the heavenly, flagging an intrinsic association between the quest for information and a more profound, extraordinary reality.

The investigation of Theorophy drives us to the acknowledgment that it isn't bound to a particular strict teaching or doctrine. All things considered, Theorophy draws motivation from the all inclusive strings that go through different otherworldly practices, recognizing the shared conviction that underlies assorted ways to the heavenly. A way of thinking rises above strict limits, embracing the enduring insight tracked down in various societies and ages.

Key to the comprehension of Theorophy is its accentuation on mysticism — the part of reasoning that dives into the idea of reality past the perceptible, actual world. Theorophy sets that the material domain isn't a definitive reality; rather, it is an indication of more profound, magical standards. This investigation includes scrutinizing the idea of awareness, oneself, and the connection between the unmanifest and manifest components of the real world.

The underlying foundations of Theorophy can be followed back to old philosophical customs, where masterminds wrestled with inquiries concerning the idea of the real world and the presence of an extreme truth past the material world. The Nonromantic and Neoplatonic schools, with their investigations into the mystical domain and the idea of the One, laid the basis for Theorophical thought. The lessons of Pythagoras, underlining the enchanted meaning of numbers, additionally reverberate with Theorophical topics of interconnectedness and all inclusive standards.

As Theorophy navigates through time, it experiences the philosophical legacy of the East, especially inside Hinduism and Buddhism. Antiquated Indian methods of reasoning, like Vedanta and Samkhya, investigate the idea of extreme reality, oneself, and the interconnectedness, everything being equal. The Upanishads, primary texts in Vedanta, dig into the idea of Brahman — a definitive, nebulous reality that underlies and binds together the variety of the manifest world.

Buddhism, with its attention on the idea of anguish and the way to edification, adds to's how Theorophy might interpret the human condition. The investigation of cognizance, temporariness, and the interconnectedness of all peculiarities in Buddhist idea lines up with Theorophical points of view on the extraordinary idea of the real world.

Theorophy's underlying foundations likewise stretch out into the supernatural customs of the Center East. Sufism, the enchanted component of Islam, gives bits of knowledge into the immediate experience of the heavenly and the excursion of the spirit toward association with the Outright.

The compositions of spiritualists like Rumi and Ibn Arabi reverberate with Theorophical subjects, underscoring affection, consideration, and the extraordinary force of profound practices.

Inside Judaism, the Kabbalistic custom offers an enchanted translation of the Torah, investigating the idea of God, the making of the universe, and the way to profound rising. The representative language of Kabbalah, with its accentuation on the Tree of Life and the interconnectedness of heavenly transmissions, lines up with Theorophical ideas of consecrated imagery and general insights.

Christian enchantment, from the compositions of the Desert Fathers to the archaic spiritualists like Meister Eckhart and Julian of Norwich, adds to the Theorophical woven artwork. The accentuation on direct fellowship with the heavenly, the extraordinary force of affection, and the association of the spirit with God resounds with Theorophical topics of inward arousing and the greatness of the egoic self.

In the East, Chinese and Daoist ways of thinking likewise leave their engraving on Theorophical thought. Daoism, with its accentuation on the Dao (the Way) and the regular progression of presence, lines up with Theorophical thoughts of concordance, interconnectedness, and the quest for a healthy lifestyle.

Theorophy recognizes the lasting way of thinking — an idea recommending that there exists a center arrangement of all inclusive bits of insight basic all strict and otherworldly practices. That's what this viewpoint sets, past the variety of social articulations, there is an ongoing idea of shrewdness that addresses the fundamental idea of the human experience and the interconnected solidarity of all presence.

As Theorophy coordinates these different strands of antiquated shrewdness, it underscores the experiential element of otherworldliness. Theorophical lessons support direct experiences with the otherworldly instead of dependence exclusively on scholarly comprehension. The spiritualist's excursion, independent of social or strict setting, is viewed as an immediate fellowship with the heavenly — a groundbreaking and binding together experience that rises above the impediments of language and creed.

Theorophy draws in with the idea of the Outright — an unutterable, extraordinary reality that fills in as the source and pith of all that exists. This Outright, whether alluded to as Brahman, the One, or by different names in different practices, is set as the constant reality in the midst of the steadily changing peculiarities of the material world. Theorophy considers the idea of this Extreme Reality and its suggestions for the comprehension of self and universe.

In the excursion of Theorophy, the idea of solidarity awareness arises. This point of view states that, at a more profound degree of the real world, everything is interconnected.

Theorophy challenges the dualistic perspective that sees partition among self and other, individual and aggregate, human and nature. The interconnectedness of all presence turns into a focal topic, cultivating a feeling of obligation and stewardship for the prosperity of the whole trap of life.

Theorophy likewise draws in with the idea of time, testing the regular straight viewpoint. It proposes that the past, present, and future are interconnected in an immortal dance. Theorophy sets the presence of an everlasting second open to the individuals who rise above customary discernment, giving a brief look into an immortal domain where all prospects coincide.

Images and models hold a huge spot in Theorophical consideration. Perceiving the groundbreaking force of images, whether tracked down in strict iconography, folklore, or dreams, Theorophy supports the consideration and translation of these images for the purpose of opening the secret insight encoded inside them. Images act as scaffolds between the limited and the boundless, the cognizant and the oblivious, offering windows into the more profound layers of the mind and the aggregate oblivious.

Theorophy recognizes the extraordinary capability of internal work and self-revelation. It perceives the interchange of light and shadow inside the individual and society. As opposed to denouncing the shadow angles, Theorophy considers them to be open doors for development and self-revelation — essential to the excursion towards completeness. Theorophical lessons underscore the significance of coordinating and rising above the dualities inborn in the human experience, encouraging a reasonable and amicable way to deal with life.

In investigating the pith of Theorophy, the idea of adoration and empathy arises as a focal subject. Love, in the Theorophical sense, isn't just a feeling yet the actual texture of presence — the binding together power that ties everything together. Theorophy calls people to encapsulate love in their viewpoints, words, and activities, perceiving that genuine otherworldliness is indistinguishable from a merciful approach to being on the planet.

Theorophy is definitely not a withdrawn, dynamic way of thinking however a down to earth guide for carrying on with a significant and intentional life. It imagines a general public wherein people, grounded in otherworldly standards, effectively add to the prosperity of humankind and the planet. The quest for material abundance is offset with a more profound comprehension of the otherworldly elements of life, encouraging an agreeable joining of material and otherworldly qualities.

Variety is a sign of Theorophical thought. Theorophy perceives that various people might resound with various ways and customs. It recognizes the variety of ways to deal with the heavenly and energizes a deferential investigation of different otherworldly roads. The quintessence of Theorophy lies in embracing this variety while perceiving the basic solidarity that rises above social and strict limits.

Theorophy draws in with the perpetual inquiry of theodicy — the test of accommodating the presence of misery and evil with the idea of a generous and supreme heavenly. It recognizes the impediments of human comprehension and supports lowliness notwithstanding the unexplored world. Theorophy welcomes people to embrace the secret as opposed to capitulating to the excessive arrogance of guaranteeing outright sureness.

In the quintessence of Theorophy, the idea of illumination takes on a nuanced and complex importance. It isn't just a scholarly seeing yet a lived insight of enlivening to one's real essence. Theorophy perceives that illumination is a continuous cycle, a persistent unfurling of higher conditions of cognizance and a developing of intelligence.

Theorophy energizes a reconsideration of the idea of self, rising above the restricted egoic personality and perceiving the more profound, general self that is interconnected with all of existence.

1.3 Setting the stage for a journey into the essence of Theorophy.

Making way for an excursion into the pith of Theorophy includes exploring the unpredictable embroidery of philosophical, religious, and magical aspects that characterize this comprehensive way to deal with figuring out presence. The expression "Theorophy" itself, got from the Greek words "theos" (god) and "philo" (love or shrewdness), indicates a significant joining of heavenly insight and human request. As we leave on this investigation, we step past the limits of customary idea, diving into a domain that rises above strict doctrine and embraces the enduring bits of insight tracked down in different otherworldly practices.

At the core of Theorophy is the acknowledgment that humankind's mission for understanding isn't restricted to the tight halls of any single conviction framework. All things being equal, Theorophy draws motivation from the general strings woven into the texture of assorted social and philosophical customs. This inclusivity is essential, recognizing the common yearnings and immortal insight that reverberation across the limits of existence.

Insightfully, Theorophy welcomes us to wrestle with supernatural inquiries concerning the idea of the real world. It provokes us to scrutinize the limits of our standard discernments and to investigate the inconspicuous aspects that underlie the manifest world. This magical request includes considering the idea of awareness, oneself, and the connection between the material and otherworldly parts of presence. Theorophy perceives that reality reaches out past what is quickly clear and urges us to leave on an excursion of self-disclosure that goes past the superficial comprehension of the world.

The underlying foundations of Theorophy venture profound into the verifiable soil of old philosophical practices. The Non-romantic and Neoplatonic ways of thinking, with their investigations into the idea of the One and the powerful components of the real world, structure a fundamental starting point for Theorophical viewpoints.

These old Greek philosophical practices give a focal point through which Theorophy examines the extraordinary parts of presence and the exchange between the noticeable and the undetectable.

Pythagoras, a figure loved both as a thinker and mathematician, adds to Theorophical thought with his investigation of the magical meaning of numbers. The idea of an extraordinary numerical request hidden the universe lines up with Theorophical subjects of interconnectedness and general standards. Pythagoras' lessons lay a basis for understanding the emblematic and mysterious language frequently utilized in Theorophical investigation.

The excursion into Theorophy additionally navigates the rich scene of Eastern way of thinking. In the antiquated Indian customs, Theorophy finds reverberation in ways of thinking, for example, Vedanta and Samkhya, which dive into the idea of extreme reality and the interconnectedness, everything being equal. The Upanishads, central texts in Vedanta, give experiences into the idea of Brahman — the amorphous, extreme reality that underlies the variety of the manifest world.

Buddhism, with its accentuation on figuring out the idea of affliction and the way to edification, adds to Theorophical experiences into the human condition. The investigation of awareness, temporariness, and the interconnectedness of all peculiarities in Buddhist idea lines up with Theorophical points of view on the otherworldly idea of the real world. Theorophy perceives the all inclusiveness of specific insights and draws from these Eastern philosophical practices to advance comprehension its might interpret the human experience.

Moving toward the west, Theorophy experiences the supernatural practices of the Center East. Sufism, the supernatural component of Islam, offers significant bits of knowledge into the immediate experience of the heavenly and the excursion of the spirit toward association with the Outright. The compositions of Sufi spiritualists like Rumi and Ibn Arabi reverberate with Theorophical topics, underscoring adoration, thought, and the groundbreaking force of otherworldly practices. Theorophy recognizes the magical aspects inside Islam and integrates these points of view into its more extensive investigation of the heavenly.

Inside Judaism, Theorophy tracks down reverberation in the Kabbalistic custom. Kabbalah gives a magical understanding of the Torah, investigating the idea of God, the production of the universe, and the way to otherworldly rising. The representative language of Kabbalah, with its accentuation on the Tree of Life and the interconnectedness of heavenly transmissions, lines up with Theorophical thoughts of hallowed imagery and all inclusive bits of insight.

Christian otherworldliness, spreading over from the compositions of the Desert Fathers to archaic spiritualists like Meister Eckhart and Julian of Norwich, adds to the Theorophical embroidery.

The accentuation on direct fellowship with the heavenly, the extraordinary force of adoration, and the association of the spirit with God reverberates with Theorophical topics of inward arousing and the greatness of the egoic self. Theorophy perceives the profundity of enchanted experiences inside Christianity and attracts from these practices to improve its investigation of the heavenly.

In the East, Chinese and Daoist ways of thinking add further layers to Theorophical thought. Daoism, with its accentuation on the Dao (the Way) and the normal progression of presence, lines up with Theorophical thoughts of congruity, interconnectedness, and the quest for a healthy lifestyle. Theorophy perceives the variety of astuteness inside Daoist reasoning and coordinates these viewpoints into its more extensive comprehension of the interaction between the human and the grandiose.

Theorophy embraces the perpetual way of thinking — an idea proposing that there exists a center arrangement of general insights fundamental all strict and profound customs. This point of view states that, past the variety of social articulations, there is a consistent idea of intelligence that addresses the fundamental idea of the human experience and the interconnected solidarity of all presence. Theorophy recognizes the worth of this perpetual insight and meshes it into its investigation of the heavenly.

As Theorophy coordinates these different strands of antiquated shrewdness, it underlines the experiential element of otherworldliness. Theorophical lessons support direct experiences with the extraordinary instead of dependence exclusively on scholarly comprehension. The spiritualist's excursion, independent of social or strict setting, is viewed as an immediate fellowship with the heavenly — an extraordinary and bringing together experience that rises above the restrictions of language and creed.

Mysticism, a central mainstay of Theorophy, welcomes people to scrutinize the idea of the real world and to investigate aspects past the material world. Theorophy places that the actual domain isn't a definitive reality; rather, it is a sign of more profound, magical standards. This investigation includes considering the idea of awareness, oneself, and the connection between the unmanifest and manifest elements of the real world. Theorophy perceives that reality stretches out past what is promptly obvious and urges people to leave on an excursion of self-revelation that goes past the superficial comprehension of the world.

Theorophy draws in with the idea of the Outright — an unutterable, otherworldly reality that fills in as the source and embodiment of all that exists. This Outright, whether alluded to as Brahman, the One, or by different names in different practices, is placed as the perpetual reality in the midst of the consistently changing peculiarities of the material world. Theorophy considers the idea of this Extreme Reality and its suggestions for the comprehension of self and universe.

Theorophy likewise stresses the idea of solidarity cognizance — a point of view that declares that, at a more profound degree of the real world, everything is interconnected. This difficulties the dualistic perspective that sees partition among self and other, individual and aggregate, human and nature. The interconnectedness of all presence turns into a focal subject, cultivating a feeling of obligation and stewardship for the prosperity of the whole trap of life.

Time, inside Theorophy, takes on an unexpected aspect in comparison to the regular direct point of view. Theorophy proposes that the past, present, and future are interconnected in an immortal dance. It sets the presence of an everlasting second open to the individuals who rise above customary insight, giving a brief look into an immortal domain where all prospects coincide. This non-straight comprehension of time has significant ramifications for how people see their own excursion and the unfurling of the real world.

Images and prime examples hold a critical spot in Theorophical examination. Theorophy perceives the extraordinary force of images, whether tracked down in strict iconography, folklore, or dreams. Images act as scaffolds between the limited and the boundless, the cognizant and the oblivious, offering windows into the more profound layers of the mind and the aggregate oblivious. Theorophy supports the consideration and understanding of these images for the purpose of opening the secret insight encoded inside them.

The groundbreaking capability of internal work and self-revelation is one more key part of Theorophy. It recognizes the transaction of light and shadow inside the individual and society. Instead of censuring the shadow viewpoints, Theorophy considers them to be valuable open doors for development and self-disclosure — vital to the excursion towards completeness. Theorophical lessons underline the significance of coordinating and rising above the dualities intrinsic in the human experience, encouraging a reasonable and agreeable way to deal with life.

In investigating the pith of Theorophy, the idea of affection and sympathy arises as a focal subject. Love, in the Theorophical sense, isn't simply a feeling yet the actual texture of presence — the bringing together power that ties everything together. Theorophy calls people to exemplify love in their thought.

The substance of Theorophy lies at the crossing point of reasoning, philosophy, and enchantment, offering an all encompassing way to deal with grasping the principal idea of presence. The actual term, got from the Greek words "theos" (god) and "philo" (love or shrewdness), exemplifies the significant thought of an affection or intelligence relating to the heavenly. It addresses an excursion into the profundities of cognizance and reality, rising above the limits of regular idea and drawing motivation from different profound customs.

At its center, Theorophy perceives that the journey for understanding reaches out past the limits of any single strict or philosophical framework. It looks to wind around together the widespread strings that go through different social and verifiable settings, recognizing the common thinking that underlies various ways to the heavenly. This inclusivity is a characterizing highlight, encouraging a feeling of solidarity and interconnectedness in the investigation of significant existential inquiries.

Thoughtfully, Theorophy welcomes people to draw in with otherworldly requests about the idea of the real world. It challenges the limits of common discernment, encouraging a more profound investigation of concealed aspects that underlie the manifest world. This otherworldly request includes considering the idea of awareness, oneself, and the multifaceted connection between the material and extraordinary parts of presence. Theorophy is a call to address and rise above the limits of customary grasping, empowering a significant investigation of the secrets that lie past the superficial perception of the world.

The foundations of Theorophy dive into old philosophical practices, where scholars wrestled with questions concerning the idea of the real world and the presence of an extreme truth past the material domain. The Non-romantic and Neoplatonic schools, with their investigations into the magical domain and the idea of the One, lay a central basis for Theorophical points of view. These old Greek customs give a focal point through which Theorophy examines the otherworldly parts of presence and the interchange between the noticeable and the imperceptible.

Pythagoras, a scholar and mathematician of classical times, adds to Theorophical thought with his investigation of the otherworldly meaning of numbers. The idea of an otherworldly numerical request hidden the universe lines up with Theorophical topics of interconnectedness and general standards. Pythagoras' lessons lay the preparation for understanding the emblematic and otherworldly language frequently utilized in Theorophical investigation.

As Theorophy travels toward the east, it experiences the rich philosophical legacy of old Indian customs. Vedanta and Samkhya ways of thinking dive into the idea of extreme reality and the interconnectedness, everything being equal. The Upanishads, primary texts in Vedanta, give significant experiences into the idea of Brahman — the nebulous, extreme reality that underlies the variety of the manifest world. Theorophy perceives the profundity of shrewdness inside these Eastern customs and integrates their bits of knowledge into its more extensive comprehension of the human experience.

Buddhism, with its attention on grasping the idea of anguish and the way to edification, contributes important experiences to Theorophy in regards to the human condition. The investigation of awareness, fleetingness, and the interconnectedness of all peculiarities in Buddhist idea lines up with Theorophical viewpoints on the extraordinary idea of the real world.

Theorophy recognizes the comprehensiveness of specific bits of insight and draws from these Eastern philosophical practices to improve its investigation of the heavenly.

Turning towards the enchanted practices of the Center East, Theorophy tracks down reverberation in Sufism — the magical component of Islam. Sufism offers significant bits of knowledge into the immediate experience of the heavenly and the excursion of the spirit toward association with the Outright. The works of Sufi spiritualists like Rumi and Ibn Arabi reverberate with Theorophical topics, accentuating affection, thought, and the extraordinary force of otherworldly practices. Theorophy recognizes the mysterious aspects inside Islam and integrates these viewpoints into its more extensive investigation of the heavenly.

Inside Judaism, Theorophy finds reverberation in the Kabbalistic custom. Kabbalah gives an enchanted understanding of the Torah, investigating the idea of God, the formation of the universe, and the way to profound climb. The representative language of Kabbalah, with its accentuation on the Tree of Life and the interconnectedness of heavenly transmissions, lines up with Theorophical ideas of holy imagery and general insights. Theorophy perceives the wealth of mysterious experiences inside Judaism and coordinates these customs into its more extensive comprehension of the heavenly.

Christian supernatural quality, crossing from the works of the Desert Fathers to archaic spiritualists like Meister Eckhart and Julian of Norwich, adds to the Theorophical embroidered artwork. The accentuation on direct fellowship with the heavenly, the groundbreaking force of affection, and the association of the spirit with God reverberates with Theorophical subjects of inward arousing and the amazing quality of the egoic self. Theorophy perceives the profundity of enchanted bits of knowledge inside Christianity and attracts from these practices to advance its investigation of the heavenly.

In the East, Chinese and Daoist methods of reasoning add further layers to Theorophical thought. Daoism, with its accentuation on the Dao (the Way) and the normal progression of presence, lines up with Theorophical thoughts of concordance, interconnectedness, and the quest for a healthy lifestyle. Theorophy perceives the variety of insight inside Daoist reasoning and coordinates these viewpoints into its more extensive comprehension of the interaction between the human and the inestimable.

Theorophy recognizes the enduring way of thinking — an idea recommending that there exists a center arrangement of widespread bits of insight fundamental all strict and profound practices. This point of view sets that, past the variety of social articulations, there is an ongoing idea of insight that addresses the fundamental idea of the human experience and the interconnected solidarity of all presence. Theorophy commends this all inclusive insight, meshing it into its investigation of the heavenly.

As Theorophy incorporates these different strands of old insight, it underlines the experiential component of otherworldliness. Theorophical lessons energize direct experiences with the extraordinary instead of dependence exclusively on scholarly comprehension. The spiritualist's excursion, independent of social or strict setting, is viewed as an immediate fellowship with the heavenly — an extraordinary and bringing together experience that rises above the restrictions of language and creed.

Transcendentalism, a primary mainstay of Theorophy, welcomes people to scrutinize the idea of the real world and to investigate aspects past the material world. Theorophy sets that the actual domain isn't a definitive reality; rather, it is an indication of more profound, mystical standards. This investigation includes pondering the idea of cognizance, oneself, and the connection between the unmanifest and manifest components of the real world. Theorophy perceives that reality reaches out past what is quickly obvious and urges people to set out on an excursion of self-revelation that goes past the superficial comprehension of the world.

Theorophy draws in with the idea of the Outright — an unspeakable, otherworldly reality that fills in as the source and embodiment of all that exists. This Outright, whether alluded to as Brahman, the One, or by different names in different customs, is set as the constant reality in the midst of the steadily changing peculiarities of the material world. Theorophy examines the idea of this Extreme Reality and its suggestions for the comprehension of self and universe.

Theorophy likewise underlines the idea of solidarity cognizance — a viewpoint that states that, at a more profound degree of the real world, everything is interconnected. This difficulties the dualistic perspective that sees detachment among self and other, individual and aggregate, human and nature. The interconnectedness of all presence turns into a focal topic, encouraging a feeling of obligation and stewardship for the prosperity of the whole trap of life.

Time, inside Theorophy, takes on an unexpected aspect in comparison to the customary direct point of view. Theorophy recommends that the past, present, and future are interconnected in an immortal dance. It places the presence of an everlasting second open to the people who rise above customary insight, giving a brief look into an immortal domain where all prospects coincide. This non-direct comprehension of time has significant ramifications for how people see their own excursion and the unfurling of the real world.

Images and prime examples hold a huge spot in Theorophical consideration. Theorophy perceives the extraordinary force of images, whether tracked down in strict iconography, folklore, or dreams. Images act as extensions between the limited and the endless, the cognizant and the oblivious, offering windows into the more profound layers of the mind and the aggregate oblivious. Theorophy energizes the examination and understanding of these images for of opening the secret insight encoded inside them.

The extraordinary capability of inward work and self-disclosure is one more key part of Theorophy. It recognizes the interaction of light and shadow inside the individual and society. Instead of censuring the shadow perspectives, Theorophy considers them to be valuable open doors for development and self-disclosure — necessary to the excursion towards completeness. Theorophical lessons stress the significance of coordinating and rising above the dualities inborn in the human experience, encouraging a fair and agreeable way to deal with life.

Chapter 2

Origins and Historical Context

The beginnings of human progress are covered in the fogs of time, darkened by the entry of centuries and the delicacy of antiquated relics. To unwind the embroidery of our previous, one should set out on an excursion through the chronicles of history, investigating the mind boggling interaction of societies, social orders, and the unyielding walk of time. The verifiable setting that approaches the starting points of human civilization is a kaleidoscope of occasions, developments, and relocations that have shaped the direction of our species.

The story starts in the far off past, in reality as we know it where agrarian social orders wandered the huge scenes, wrestling with the difficulties of endurance. The roaming way of life of these early people was directed by the rhythms of nature, as they navigated scenes looking for food. The simple devices designed from stone and bone were the signs of this age, and the dominance of fire denoted a significant second in the rising of Homo sapiens.

As time streamed like a waterway through the ages, mankind went through an extraordinary excursion from roaming ways of life to settled rural networks. The Neolithic Upset, an essential part in mankind's set of experiences, saw the development of farming as a progressive power. The taming of plants and creatures proclaimed a seismic shift, changing the actual texture of human life. Settlements prospered, and the once-meandering clans started to develop the land, planting the seeds of human advancement.

The Fruitful Bow, a district reaching out from the Nile Valley to the Tigris and Euphrates, turned into the support of progress. Mesopotamia, with its ripe soils, saw the ascent of strong city-states like Sumer and Akkad. These antiquated social orders laid the foundation for the improvement of complicated social designs, administration frameworks, and the earliest known types of composing.

All the while, along the banks of the Nile, the old Egyptians developed a civilization that persevered for centuries. The stupendous designs of the pyramids, the Sphinx, and the rambling urban communities along the Nile validated the magnificence of this antiquated civilization. The Pharaohs controlled with divine power, and the mind boggling strict convictions of the Egyptians saturated each feature of day to day existence.

In the Indian subcontinent, the Indus Valley Civilization prospered along the banks of the Indus Waterway. Harappa and Mohenjo-Daro were among the refined metropolitan communities that flaunted progressed metropolitan preparation, unpredictable waste frameworks, and a composing framework yet to be completely interpreted.

The many-sided curios uncovered from the remnants of these urban communities offer tempting looks into the existences of these antiquated people groups.

Far toward the east, the old Chinese human advancement developed along the Yellow Stream. The Xia, Shang, and Zhou administrations stamped progressive stages in the advancement of Chinese culture, establishing the groundwork for philosophical customs, administration frameworks, and mechanical developments. The Incomparable Wall, a demonstration of human creativity and determination, arose as a considerable hindrance against outer dangers.

The Mediterranean bowl, a junction of civilizations, saw the ascent of old Greece. The city-provinces of Athens and Sparta became focal points of workmanship, reasoning, and political trial and error. The vote based beliefs of Athens, as expressed by scholars like Plato and Aristotle, made a permanent imprint on the course of Western idea.

As the pages of history turned, the Roman Republic arose as a prevailing power in the old world. The development of the Roman Domain introduced a time of uncommon regional successes, bringing different societies under the umbrella of Roman rule. The Pax Romana, a time of relative harmony, worked with the trading of thoughts, exchange, and social cross-fertilization.

The cauldron of history additionally gave testimony regarding the ascent of realms in the East. The Maurya and Gupta domains in India encouraged social and scholarly accomplishments, remembering the idea of zero for arithmetic and compositions on medication and cosmology. In Persia, the Achaemenid Domain under Cyrus the Incomparable made an immense and different domain that extended from the Mediterranean to the Indus.

Nonetheless, the tides of history are whimsical, and the magnificence of realms frequently respects the unyielding powers of progress. The fall of the Western Roman Realm in 476 CE denoted a turning point ever, introducing the wild time frame known as the Medieval times. Europe, dove into a mess, saw the fracture of political power, the ascent of feudalism, and the domination of the Congregation as a prevailing organization.

In the midst of the shadows of the middle age time frame, a social and scholarly recovery known as the Renaissance bloomed in Europe. The rediscovery of traditional texts, the investigation of new skylines, and the blooming of creative articulation characterized this time of resurrection. Any semblance of Leonardo da Vinci, Michelangelo, and Raphael made a permanent imprint on the material of human accomplishment.

However, as the pendulum of history swung, so too did the undeniable trends blow across the mainlands. In the East, the Islamic Brilliant Age turned into a signal of illumination, encouraging headways in science, medication, reasoning, and human expression.

The Place of Shrewdness in Baghdad turned into a cauldron of information, where researchers from different foundations united to decipher and combine the insight of old civic establishments.

The period of investigation opened new sections in the story of mankind's set of experiences. Journeys of disclosure, drove by figures like Christopher Columbus, Ferdinand Magellan, and Vasco da Gama, extended the explored parts of the planet and fashioned worldwide associations. The Columbian Trade, a mind boggling trap of natural, social, and financial trades between the Old World and the New, reshaped social orders on the two sides of the Atlantic.

The beginning of the cutting edge time saw seismic changes in the political, financial, and social scenes. The Renaissance gave way to the Illumination, a scholarly development that advocated reason, individual freedoms, and the quest for information. Masterminds like John Locke, Voltaire, and Montesquieu laid the foundation for the beliefs that would shape the popularity based investigations of the next few centuries.

The Modern Upset, an extraordinary period set apart by mechanical developments and the motorization of creation, introduced another time of private enterprise and urbanization. Plants burped smoke high up, and steam motors fueled trains that navigated mainlands. The monetary and social texture of social orders went through significant changes, as the agrarian lifestyle respected the objectives of modern advancement.

The expanding influences of these progressions resonated across the globe, igniting developments for political autonomy, civil rights, and equity. The American Insurgency, with its call forever, freedom, and the quest for joy, set up for the development of republics and sacred vote based systems. The French Unrest, a cauldron of revolutionary goals, looked to destroy the remnants of feudalism and gentry.

In the nineteenth hundred years, the enthusiasm for patriotism cleared across Europe and then some, reshaping political limits and personalities. The unification of Germany and Italy and the disintegration of realms in Focal and Eastern Europe mirrored the tide of patriot yearnings. All the while, majestic powers took part in a race for settlements, cutting up Africa, Asia, and the Pacific in a scramble for assets and international impact.

The twentieth century unfurled as a pot of extraordinary difficulties and valuable open doors. The two Universal Conflicts, decimating in scale and effect, reshaped the international scene and redrew public boundaries. The ascent of extremist systems, exemplified by Adolf Hitler's Nazi Germany and Joseph Stalin's Soviet Association, cast dim shadows over the goals of a majority rule government and basic liberties.

The post-war period saw the development of a bipolar world request, with the US and the Soviet Association secured in a Virus War.

The philosophical and international battle among private enterprise and socialism worked out on a worldwide stage, molding unions, clashes, and the direction of countries. The ghost of atomic weapons cast a long shadow, upping the ante of global relations to exceptional levels.

As the Virus War defrosted in the last 50% of the twentieth hundred years, new elements arose on the world stage. The appearance of globalization introduced a time of interconnectedness, as mechanical advances in correspondence and transportation spanned the holes among societies and economies. The ascent of worldwide enterprises, the expansion of the web, and the speed increase of exchange and money reshaped the forms of the worldwide scene.

The breakdown of the Soviet Association in 1991 denoted the finish of the Virus War and proclaimed a unipolar second, with the US remaining as the transcendent worldwide power. Nonetheless, the 21st century has seen the development of new international real factors and difficulties. The ascent of China as a financial force to be reckoned with, the reassertion of Russia on the worldwide stage, and the moving sands of unions and clashes mirror the intricate and dynamic nature of our interconnected world.

All the while, the 21st century has wrestled with phenomenal worldwide difficulties. Environmental change, an approaching existential danger, has incited dire calls for aggregate activity to moderate its effects. The multiplication of innovation, while offering extraordinary potential outcomes, has raised moral and social quandaries that request smart thought. The continuous mission for civil rights, uniformity, and common freedoms stays a focal and incomplete account in the tale of humankind.

2.1 Tracing the historical development of Theorophy through various cultures and civilizations.

The investigation of the verifiable improvement of Theosophy takes us on an intriguing excursion through different societies and developments, each contributing remarkable strings to the rich embroidery of this recondite custom. Theosophy, got from the Greek words "theos" (importance divine) and "sophia" (meaning insight), is a magical and philosophical framework that looks to disentangle the secrets of presence, the idea of heavenliness, and the interconnectedness of all life.

The foundations of Theosophy broaden profound into the records of antiquated astuteness, drawing motivation from the exclusive lessons of various societies. In the East, the Vedic customs of old India give a primary impact. The Upanishads, an assortment of philosophical texts that arose around the sixth century BCE, dive into the idea of the real world, oneself (atman), and a definitive reality (Brahman). Ideas like karma, resurrection, and the interconnectedness of all living creatures track down reverberation in Theosophical idea.

Also, the supernatural practices of Hinduism, especially those encapsulated in the Bhagavad Gita, the Yoga Sutras of Patanjali, and the lessons of different sages and yogis, add to the magical underpinnings of Theosophy. The possibility of a profound way prompting self-acknowledgment and association with the heavenly is an ongoing idea that winds through both old Eastern methods of reasoning and Theosophical lessons.

Moving toward the west, the Greek time frame in old Greece fills in as one more cauldron for the advancement of Theosophy. The otherworldly practices of Neoplatonism, exemplified by figures like Plotinus, tried to accommodate Non-romantic way of thinking with components of Eastern otherworldliness. The idea of an extreme, unspeakable reality from which all radiates, as well as the rising of the spirit toward divine solidarity, resounds with Theosophical topics.

The impact of Hermeticism, a recondite custom established in the lessons ascribed to Hermes Trismegistus, likewise penetrates the advancement of Theosophy. Hermeticism upholds the possibility of an otherworldly insight that uncovers the heavenly standards hidden the universe. The popular Airtight saying, "As above, so underneath," exemplifies the thought of the microcosm mirroring the world, an idea that reverberations all through Theosophical writing.

In the middle age period, during the blossoming of Islamic progress, Sufism arose as an enchanted aspect inside Islam that imparts affinities to Theosophical standards. Sufi spiritualists, through practices, for example, dhikr (recognition of God), looked to achieve direct profound experience and association with the heavenly. The accentuation on inward information, the extraordinary force of adoration, and the acknowledgment of a widespread otherworldly truth line up with key precepts of Theosophy.

The Renaissance in Europe saw a recovery of interest in old insight, prompting the rediscovery of airtight and catalytic texts. Figures like Marsilio Ficino and Giovanni Pico della Mirandola investigated the combination of Dispassionate, Airtight, and Kabbalistic thoughts, laying the preparation for an otherworldly restoration that would impact resulting exclusive developments, including Theosophy.

The nineteenth century denoted a vital second in the formalization of Theosophical lessons with the establishing of the Theosophical Society by Helena Petrovna Blavatsky, Henry Steel Olcott, and William Quan Judge. Blavatsky, a Russian soothsayer and spiritualist, assumed a focal part in combining different exclusive customs into a strong framework. Her original work, "The Mysterious Tenet," turned into a foundation of Theosophical writing.

"The Mysterious Precept" digs into cosmogony, human studies, and the otherworldly development of mankind, drawing on old sacred texts, mysterious customs, and Blavatsky's guaranteed contact with profound creatures known as the Experts of Shrewdness. The book sets the presence of an immortal and widespread insight, open through otherworldly knowledge and inward change, and it investigates the interconnectedness of all life.

Blavatsky's Theosophy underscores the possibility of otherworldly development, recommending that mankind goes through a cyclic course of improvement and refinement. The idea of root races, addressing various phases of human development, is presented, and the theosophical perspective imagines an intentional advancement directed by profound insights.

The Theosophical Society, with its base camp at first in New York and later in Adyar, India, turned into a cauldron for the dispersal of recondite lessons. The association pulled in a different exhibit of people, including erudite people, specialists, and otherworldly searchers. Annie Besant, a noticeable figure in the Theosophical development, succeeded Blavatsky and assumed a key part in promoting Theosophy, especially in India.

Besant's contribution in the Indian patriot development and her relationship with Jiddu Krishnamurti, a youthful Indian kid trusted by some to be the "World Educator" or the vehicle for another profound regulation, added layers of intricacy to the Theosophical story. The Request for the Star in the East, framed to set up the world for Krishnamurti's expected job, mirrored the eschatological and groundbreaking aspects inside Theosophy.

The twentieth century saw the proceeded with advancement of Theosophical idea, with different associations and people adding to its turn of events. Alice Bailey, a conspicuous Theosophist, established the Obscure School and worked with the Tibetan Expert Djwhal Khul to create a progression of recondite lessons. Bailey's works, underlining profound brain science, the Seven Beams, and the job of mankind in the vast arrangement, developed Theosophical standards.

The post-Blavatsky time additionally saw the foundation of the Unified Cabin of Theosophists (ULT), an association committed to the dispersal of unique Theosophical lessons without formal doctrine or incorporated power. The ULT accentuates the significance of self-study, individual utilization of standards, and the sustaining of an inward profound life.

Lined up with these turns of events, the impact of Theosophy spread all around the world, motivating craftsmen, essayists, and scholars. Crafted by Wassily Kandinsky, Piet Mondrian, and different craftsmen related with the Theosophical Society mirror the effect of elusive thoughts on creative articulation. In writing, the impact of Theosophy is obvious in progress of writers like D.H. Lawrence, W.B. Yeats, and Aldous Huxley.

The Theosophical development likewise assumed a part in the rise of the New Age development in the last 50% of the twentieth hundred years. The New Age, portrayed by a syncretic mix of otherworldly and supernatural thoughts, drew motivation from Theosophy's accentuation on profound development, exclusive insight, and the interconnectedness of all life. Ideas, for example, chakras, qualitys, and elective recuperating modalities tracked down reverberation inside this more extensive social development.

As we explore the 21st 100 years, the tradition of Theosophy perseveres, proceeding to motivate searchers on the profound way. Theosophical thoughts, with their accentuation on general fraternity, otherworldly change, and the mission for internal information, stay important during a time set apart by social variety, mechanical advances, and worldwide interconnectivity.

In following the verifiable improvement of Theosophy through different societies and civilizations, we experience an embroidery woven with strings of old insight, magical customs, and the imaginative blend of obscure bits of knowledge. From the Vedic sages of old India to the Renaissance rationalists of Europe, from the Sufi spiritualists to the Theosophical trailblazers of the nineteenth 100 years, the excursion of Theosophy mirrors an enduring journey for figuring out the secrets of presence and the idea of the heavenly.

The Theosophical practice, in its different signs, welcomes people to leave on an excursion of self-disclosure, internal change, and the acknowledgment of a common otherworldly legacy. As the torchbearers of Theosophy give it to people in the future, the investigation of these immortal lessons keeps on enlightening the way toward a more profound comprehension of the interconnected snare of life and the endless conceivable outcomes that lie not too far off of human development.

2.2 Examining key figures and texts that have contributed to the evolution of Theorophy.

To comprehend the advancement of Theosophy, one should dive into the lives and lessons of key figures who have formed this recondite practice. These people, frequently spiritualists, savants, and searchers of significant otherworldly insights, have made a permanent imprint on the turn of events and scattering of Theosophical thoughts. Inspecting their commitments gives experiences into the assorted impacts that have united to frame the complex embroidered artwork of Theosophy.

Helena Petrovna Blavatsky, a focal figure throughout the entire existence of Theosophy, remains as a light whose lessons established the groundwork for the cutting edge Theosophical development. Brought into the world in Russia in 1831, Blavatsky had an exceptional existence as a valiant explorer, essayist, and medium. Her original work, "The Mysterious Regulation," distributed in 1888, fills in as a foundation of Theosophical writing. In this stupendous work, Blavatsky dives into cosmogony, human studies, and the profound development of mankind, drawing from a huge range of sources, including old sacred writings, elusive customs, and her own guaranteed contact with otherworldly elements known as the Bosses of Shrewdness.

Blavatsky's Theosophy stresses the possibility of a lasting insight that rises above social and verifiable limits. She affirmed that this ageless insight, open through otherworldly knowledge and internal change, frames the center of exclusive lessons across various customs.

Blavatsky's mixed combination of Eastern and Western supernatural quality, Hermeticism, and old methods of reasoning laid the foundation for an all encompassing and comprehensive way to deal with otherworldly request inside Theosophy.

Henry Steel Olcott, a prime supporter of the Theosophical Society close by Blavatsky, assumed an essential part in the early turn of events and hierarchical parts of the development. Olcott, brought into the world in the US in 1832, was a complex person with interests going from regulation and farming to profound way of thinking. His obligation to the standards of fraternity, strict resilience, and the investigation of near religion became vital to the Theosophical Society's main goal. Olcott's commitments stretched out past the hypothetical domain; he effectively participated in friendly and instructive drives, especially in India, where he looked to connect social partitions and advance grasping among East and West.

William Quan Judge, another fellow benefactor of the Theosophical Society, was an Irish-American lawyer and spiritualist who assumed a huge part in the early long periods of the development. Judge, brought into the world in 1851, turned into a gave follower of Blavatsky and worked intimately with her to scatter Theosophical lessons. His legitimate foundation and sharp keenness made him a significant resource in articulating and making sense of Theosophical standards. After Blavatsky's passing, Judge kept on being a conspicuous figure in the Theosophical development, driving the American part of the general public and adding to the writing with works, for example, "The Expanse of Theosophy."

Annie Besant, a dynamic and powerful Theosophist, succeeded Blavatsky as the head of the Theosophical Society. Brought into the world in Britain in 1847, Besant's excursion into Theosophy was set apart by her scholarly ability and social activism. A vocal backer for ladies' freedoms and laborers' privileges, Besant carried an energetic and logical aspect to Theosophical lessons. Her contribution in the Indian patriot development and cooperation with Jiddu Krishnamurti added political and eschato-logical aspects to Theosophy. Besant's administration left a getting through influence on the worldwide spread of Theosophical thoughts, especially in India, where she assumed a vital part in the renaissance of Hindu way of thinking and the advancement of training.

Jiddu Krishnamurti, a puzzling and compelling figure related with Theosophy, was found by Annie Besant and C.W. Leadbeater in India in the mid twentieth 100 years. Brought into the world in 1895, Krishnamurti was broadcasted as the expected "World Educator" or the vehicle for another otherworldly regulation. The Request for the Star in the East was framed to set up the world for his normal job. Notwith-standing, in an earth shattering new development in 1929, Krishnamurti disbanded the Request and broke up the messianic assumptions encompassing him. He set out on a free way, pushing for an extreme type of profound request that rose above strict doctrine and hierarchical designs. While Krishnamurti limited any association with formal Theosophical associations, his effect on otherworldly idea and the investigation of awareness stays significant.

Alice Bailey, a critical figure in the post-Blavatsky period of Theosophy, made prom-inent commitments to obscure writing and profound way of thinking. Brought into the world in Britain in 1880, Bailey established the Little known School and worked intimately with the Tibetan Expert Djwhal Khul to create a progression of exclusive lessons. Her productive compositions, which incorporate works like "A Composition on Inestimable Fire" and "The Externalization of the Order," developed Theosophical standards. Bailey's attention on otherworldly brain research, the Seven Beams, and the job of humankind in the astronomical arrangement further advanced the embroidered artwork of Theosophical idea.

These key figures, among others, have essentially impacted the development of Theosophy through their compositions, lessons, and hierarchical endeavors. In any case, it's critical to perceive that Theosophy isn't exclusively characterized by these people; rather, it is a developing practice formed by a different exhibit of supporters. Theosophical thoughts have resounded with scholars, craftsmen, and otherworldly searchers across various social and verifiable settings, making a dynamic and complex development.

Past individual figures, certain key texts play had a urgent impact in profoundly shaping Theosophical idea and scattering its standards. As referenced before, Helena Blavatsky's "The Mysterious Regulation" remains as a perfect work of art, giving a far reaching blend of exclusive lessons. The significant bits of knowledge into cosmogony, humanities, and the profound advancement of mankind introduced in this work have been key to the improvement of Theosophical way of thinking.

Blavatsky's previous work, "Isis Disclosed," fills in as another essential message that investigates different supernatural and philosophical customs, disentangling the secret strings that interface mankind's profound legacy. Her compositions, which likewise incorporate articles, letters, and expositions, all in all structure a collection of work that keeps on being contemplated and deciphered inside the Theosophical custom.

Annie Besant's works, for example, "The Antiquated Insight" and "Thought Power: Its Control and Culture," added to the advancement and elucidation of Theosophical thoughts. Besant's capacity to explain complex powerful ideas in open language spoke to a wide crowd and assumed a pivotal part in the worldwide spread of Theosophy.

William Judge's "The Expanse of Theosophy" stays an original work that gives a reasonable and succinct outline of Theosophical standards. Composed as a progression of examples, the book offers a methodical investigation of key ideas, including karma, rebirth, and the idea of the heavenly. Judge's commitment to Theosophical writing has been instrumental in acquainting rookies with the basic fundamentals of the practice.

Alice Bailey's series of books, on the whole known as "The Alice A. Bailey Books," comprise a critical group of work that expands upon Theosophical standards and presents new ideas. The emphasis on the Seven Beams, otherworldly brain research, and the combination of different elusive customs mirrors Bailey's special commitment to the advancing scene of Theosophy.

Notwithstanding these essential messages, different optional works, analyses, and understandings by Theosophists and researchers have additionally enhanced the writing of Theosophy. The advancing idea of Theosophical idea is apparent in the consistent investigation and reevaluation of essential messages by progressive ages of searchers and researchers.

The Theosophical development has additionally been set apart by authoritative turns of events and the rise of different Theosophical social orders and schools. While the first Theosophical Society established by Blavatsky, Olcott, and Judge stays persuasive, different associations have shaped, each with its own accentuation and understanding of Theosophical standards. The Assembled Cabin of Theosophists (ULT), established by Katherine Tingley, separates itself by pushing for the scattering of unique Theosophical lessons without formal authoritative opinion or concentrated power.

The verifiable advancement of Theosophy has not been without difficulties, discussions, and inside divisions. Disagreements regarding authority, understandings of key lessons, and dissimilar authoritative designs have denoted the historical backdrop of the Theosophical development. In any case, this dynamism and variety inside the development mirror the living idea of Theosophy as an advancing profound practice.

As Theosophy keeps on developing in the 21st 100 years, it faces the difficulties and potential open doors introduced by a quickly impacting world. The worldwide interconnectedness worked with by innovation, the resurgence of interest in other-worldliness and mystery, and the basic for natural manageability are factors that shape the contemporary scene in which Theosophy tracks down itself.

2.3 Highlighting the cross-cultural influences that have shaped Theorophical thought.

The development of Theosophical idea is set apart by a captivating transaction of multifaceted impacts, as this exclusive custom draws from different enchanted, philosophical, and profound practices across various landmasses and ages. The investigation of these culturally diverse impacts gives experiences into the rich embroidery of Theosophy, uncovering how different strands of intelligence have woven together to frame a comprehensive and comprehensive way to deal with profound request.

One of the basic impacts on Theosophical idea is established in the antiquated otherworldly practices of India. Helena Petrovna Blavatsky, a focal figure in the improvement of Theosophy, broadly attracted upon Hindu and Vedantic methods of reasoning forming the center standards of the development.

The Upanishads, with their significant investigations of the idea of the real world, oneself (atman), and a definitive reality (Brahman), essentially formed's comprehension Blavatsky might interpret the magical underpinnings of Theosophy. Ideas like karma, resurrection, and the interconnectedness of every single living being, basic to Theosophical idea, track down reverberation in these old Indian lessons.

Besides, Blavatsky's commitment with the Bhagavad Gita, a fundamental text inside the Indian legendary Mahabharata, assumed a crucial part in forming Theosophical thoughts. The Gita's investigation of obligation (dharma), the way of caring activity (karma yoga), and the quest for otherworldly information line up with key fundamentals of Theosophy, accentuating the significance of moral living, self-change, and the journey for higher insight.

Moving past India, Theosophical idea has been altogether impacted by components of Buddhism, especially Mahayana Buddhism. The accentuation on empathy, the interconnectedness of all life, and the Bodhisattva ideal — a being committed to the illumination and government assistance of others — all reverberate with Theosophical standards. Blavatsky's investigation of Tibetan Buddhism and her presentation of the idea of the Tibetan Bosses of Intelligence mirror the multifaceted flows among Theosophy and Tibetan profound practices.

Old Egyptian magic likewise added to the diverse blend of impacts shaping Theosophical idea. The Egyptian Book of the Dead and the mysterious lessons related with the old Egyptian secret schools gave a background to Blavatsky's investigation of recondite information. Images, for example, the ankh and the Eye of Horus, tracked down their direction into Theosophical iconography, addressing aspects of other-worldly importance and antiquated intelligence.

The Greek time frame in old Greece fills in as one more wellspring of culturally diverse effects on Theosophy. The mysterious customs of Neoplatonism, led by rationalists like Plotinus, tried to accommodate Non-romantic way of thinking with components of Eastern supernatural quality. The idea of an unutterable reality from which all radiates and the rising of the spirit toward divine solidarity resound with Theosophical subjects. The impact of Hermeticism, an elusive custom with establishes in Greek Egypt, likewise saturates Theosophical idea, as reflected in the Airtight saying "As above, so beneath."

The Jewish mysterious practice of Kabbalah contributes one more layer to the culturally diverse embroidered artwork of Theosophy. Ideas inside Kabbalah, like the Tree of Life and the radiations of the heavenly (sefirot), bear charming equals to Theosophical cosmology. The syncretic methodology of Theosophy, drawing upon assorted profound customs, mirrors an eagerness to track down shared traits and shared bits of knowledge across various societies.

Moving into the middle age time frame, the Sufi spiritualists inside Islamic progress assumed a critical part in shaping Theosophical idea. Sufism, with its accentuation on the immediate experience of the heavenly, the groundbreaking force of adoration (ishq), and the acknowledgment of a general profound truth, reverberates with Theosophical standards. Blavatsky's experiences with Sufi otherworldliness during her movements in the East left an engraving on how she might interpret the magical components of Islam.

The Renaissance in Europe denoted a resurgence of interest in old insight, prompting the rediscovery of airtight and catalytic texts. Crafted by Marsilio Ficino and Giovanni Pico della Mirandola, who tried to combine Non-romantic, Airtight, and Kabbalistic thoughts, gave a scaffold between the old and the cutting edge. The Renaissance's impact on Theosophy is apparent in its comprehensive methodology, drawing motivation from assorted social and authentic sources.

As Theosophy unfurled in the late nineteenth and mid twentieth hundreds of years, it experienced the profound customs of the native people groups of the Americas. The acknowledgment of the holiness of nature, the interconnectedness of every single living being, and the quest for otherworldly information inside Local American and First Countries societies reverberated with Theosophical standards. The Theosophical Society's commitment with these practices mirrored a receptiveness to gaining from assorted social articulations of otherworldliness.

Annie Besant, a conspicuous Theosophist who succeeded Blavatsky as the head of the Theosophical Society, assumed a key part in carrying Theosophy to India. Her profound association in Indian culture, her promotion for Indian freedom, and her coordinated effort with Indian pioneers added to the diverse trade among Theosophy and the otherworldly legacy of India. The union of Theosophical thoughts with Hindu methods of reasoning and the advancement of the Indian renaissance became vital to the Theosophical story in India.

The Theosophical development likewise saw the incorporation of Eastern and Western obscure practices. The lessons of the Tibetan Expert Djwhal Khul, as diverted by Alice Bailey, brought components of Eastern elusiveness into Theosophical idea. Bailey's emphasis on the Seven Beams, otherworldly brain research, and the union of different obscure practices mirrors a development of Theosophical thoughts because of a changing worldwide scene.

Jiddu Krishnamurti, however he reduced most, if not all, connection with formal Theosophical associations, epitomized multifaceted impacts inside the development. Found in India, Krishnamurti was related with Theosophy as the expected "World Educator." His resulting disbanding of the Request for the Star in the East and his free otherworldly request mirrored an impact of Eastern ways of thinking, especially those underlining direct private experience and the amazing quality of hierarchical designs.

The diverse impacts inside Theosophy additionally appeared in the development's commitment with Western mysterious and profound developments. The Airtight Request of the Brilliant First light, a powerful mysterious society in the late nineteenth 100 years, imparted normal subjects to Theosophy, including the investigation of the otherworldly components of Christianity, the investigation of Kabbalah, and the quest for profound commencement. The trading of thoughts among Theosophy and the Brilliant Day break embodies the penetrability of limits between various elusive practices.

In the domain of workmanship, Theosophy found articulation through craftsmen who were roused by culturally diverse profound subjects. Wassily Kandinsky, a Russian painter and unmistakable Theosophist, investigated the profound elements of workmanship and variety hypothesis. His work, especially the series "Concerning the Profound in Craftsmanship," reflects Theosophical thoughts regarding the amazing quality of realism and the statement of inward otherworldly real factors through imaginative creation.

As Theosophy unfurled across different social scenes, it turned into a gathering ground for people from various foundations — Eastern and Western, old and present day, native and pilgrim. The Theosophical Society itself, with its accentuation on general fellowship and the investigation of similar religion, epitomized an ethos of inclusivity that supported the investigation and joining of diverse impacts.

The diverse flows inside Theosophy, be that as it may, were not without challenges. The assignment of specific components from various practices and the potential for social errors highlighted the intricacies innate in culturally diverse trade. The discourse among Theosophy and different otherworldly practices likewise confronted analysis, with allegations of social appointment and error.

In exploring these difficulties, Theosophy has looked to cultivate a genuine and deferential commitment with different social and profound articulations. The accentuation on the all inclusiveness of otherworldly standards, the acknowledgment of a typical wellspring of intelligence across customs, and the advancement of common comprehension have been key to Theosophical endeavors to connect social partitions.

The contemporary scene of Theosophy keeps on mirroring its multifaceted impacts. The worldwide interconnectedness worked with by innovation has empowered Theosophists from various regions of the planet to participate in significant discourse and cooperation. The Theosophical Society, with its worldwide branches and different participation, embodies a proceeding with obligation to diverse trade and the investigation of shared otherworldly experiences.

As Theosophy pushes ahead into the 21st hundred years, the multifaceted impacts that have molded its idea stay essential to its character. The continuous exchange among East and West, old and current, and different profound practices improves the dynamic and developing nature of Theosophical request. In a world set apart by social variety, interconnectivity, and a developing consciousness of the worldwide legacy of otherworldly insight, Theosophy remains as a demonstration of the conceivable outcomes that arise when people and networks draw in with the extraordinary insights that join humankind across social and transient limits.

The development of Theosophical idea is an embroidery woven with strings from different magical, philosophical, and otherworldly practices, each adding to the rich and extensive perspective that describes Theosophy. These molding impacts have made a permanent imprint on the development, mirroring a union of old insight, multifaceted trades, and the visionary experiences of key figures. As we explore the unpredictable territory of Theosophical idea, it becomes obvious that its establishments are well established in a junction of different social and scholarly flows.

Old Eastern Insight:

One of the basic mainstays of Theosophical idea lays on the antiquated profound practices of the East, especially India. Helena Petrovna Blavatsky, a focal figure in the improvement of Theosophy, drew broadly from Hindu and Vedantic methods of reasoning. The Upanishads, with their significant investigations of the idea of the real world, oneself, and a definitive reality, gave a philosophical bedrock to Theosophical standards. The ideas of karma, resurrection, and the interconnectedness of every single living being, characteristic for Theosophy, found reverberations in these old Indian lessons. The Bhagavad Gita, a loved text inside Hinduism, likewise assumed a crucial part in molding Theosophical thoughts, impacting its viewpoints working, magnanimous activity, and the quest for profound information.

Buddhist Commitments:

Theosophical idea resounds with components of Mahayana Buddhism, particularly in its accentuation on empathy, the interconnectedness of all life, and the Bodhisattva ideal. These Buddhist ideas line up with key Theosophical standards, mirroring a diverse trade that improves the two customs. The investigation of Tibetan Buddhism, with its obscure aspects and accentuation on otherworldly development, further extended the interlacing of Buddhist impacts into Theosophical idea.

Greek Enchantment:

The Greek time frame in antiquated Greece denoted one more persuasive stream of felt that formed Theosophy. Neoplatonism, advocated by scholars like Plotinus, looked to accommodate Dispassionate way of thinking with components of Eastern otherworldliness. The Neoplatonic idea of an inexpressible reality from which all exudes and the climb of the spirit toward divine solidarity resounds with Theosophical subjects. The Airtight practice, established in Greek Egypt, additionally made a permanent imprint on Theosophy. The maxim "As above, so underneath," embodying the possibility of microcosm reflecting world, turned into a basic guideline inside Theosophical idea.

Kabbalistic Impacts:

Jewish magic, especially the Kabbalah, added to the syncretic idea of Theosophical idea. Ideas inside Kabbalah, like the Tree of Life and the sefirot (transmissions of the heavenly), tracked down matches in Theosophical cosmology. This diverse trade among Theosophy and Kabbalah added layers of representative extravagance to the development.

Islamic Otherworldliness (Sufism):

During her movements in the East, Blavatsky experienced Sufi enchantment inside Islamic human progress. Sufism, with its accentuation on the immediate experience of the heavenly, groundbreaking adoration (ishq), and the acknowledgment of a widespread otherworldly truth, reverberated with Theosophical standards. This experience added a mysterious aspect to Theosophy and mirrored the development's receptivity to different profound articulations.

Antiquated Egyptian Supernatural quality:

The mysterious lessons of old Egypt, remembering those found for the Egyptian Book of the Dead and inside the setting of old secret schools, gave a scenery to Theosophical investigation. Images, for example, the ankh and the Eye of Horus tracked down their direction into Theosophical iconography, representing features of otherworldly importance and antiquated shrewdness.

Native Insight:

As Theosophy unfurled, it experienced the profound practices of native people groups, especially the Local American and First Countries societies. The acknowledgment of the holiness of nature, the interconnectedness of every single living being, and the quest for profound information inside these practices reverberated with Theosophical standards. This commitment mirrored a transparency inside Theosophy to gain from and regard different social articulations of otherworldliness.

The Renaissance and Western Exclusiveness:

The Renaissance in Europe denoted a resurgence of interest in old insight, prompting the rediscovery of airtight and catalytic texts. Crafted by Renaissance savants like Marsilio Ficino and Giovanni Pico della Mirandola, who tried to integrate Nonromantic, Airtight, and Kabbalistic thoughts, gave a scaffold between the old and the cutting edge. This syncretic methodology affected Theosophical idea, which likewise drew motivation from different social and authentic sources.

Airtight Request of the Brilliant Sunrise:

The late nineteenth century saw the impact of the Airtight Request of the Brilliant Sunrise, a mysterious society that common normal topics with Theosophy. The two developments investigated the mysterious elements of Christianity, dove into the investigation of Kabbalah, and looked for profound inception. The trading of thoughts among Theosophy and the Brilliant Sunrise exemplified the penetrability of limits between various exclusive customs.

Annie Besant and the Indian Renaissance:

Annie Besant, who succeeded Blavatsky as the head of the Theosophical Society, assumed a significant part in carrying Theosophy to India. Her profound association in Indian culture, her support for Indian autonomy, and her coordinated effort with Indian pioneers added to the multifaceted trade among Theosophy and the otherworldly legacy of India. The blend of Theosophical thoughts with Hindu ways of thinking and the advancement of the Indian renaissance became vital to the Theosophical account in India.

Alice Bailey and Eastern Exclusiveness:

In the post-Blavatsky time, Alice Bailey's coordinated effort with the Tibetan Expert Djwhal Khul brought components of Eastern exclusiveness into Theosophical idea. Bailey's emphasis on the Seven Beams, profound brain research, and the combination of different recondite customs mirrors a development of Theosophical thoughts in light of a changing worldwide scene.

Jiddu Krishnamurti's Free Way:

While Jiddu Krishnamurti moved away from formal Theosophical associations, his relationship with Theosophy, especially as the expected "World Instructor," mirrored an impact of Eastern ways of thinking. Krishnamurti's free profound request, stressing direct private experience and the greatness of hierarchical designs, denoted a takeoff from conventional Theosophical methodologies and typified diverse impacts.

Theosophy and Worldwide Workmanship:

Theosophical thoughts likewise tracked down articulation in the domain of craftsmanship. Wassily Kandinsky, a Russian painter and noticeable Theosophist, investigated the profound components of craftsmanship and variety hypothesis. His work, especially the series "Concerning the Otherworldly in Craftsmanship," reflects

Theosophical thoughts regarding the greatness of realism and the declaration of inward profound real factors through imaginative creation.

As Theosophy keeps on advancing in the 21st 100 years, the culturally diverse impacts that have formed its idea stay essential to its personality. The worldwide interconnectedness worked with by innovation has empowered Theosophists from various regions of the planet to take part in significant exchange and cooperation. The Theosophical Society, with its worldwide branches and different enrollment, represents a proceeding with obligation to diverse trade and the investigation of shared profound bits of knowledge.

Theosophy's receptiveness to different social and profound articulations lines up with its central standards of general fellowship and the investigation of similar religion. In a world set apart by social variety, interconnectivity, and a developing consciousness of the worldwide legacy of profound insight, Theosophy remains as a demonstration of the potential outcomes that arise when people and networks draw in with the extraordinary bits of insight that join humankind across social and fleeting limits.

Difficulties and Reactions:

In any case, this multifaceted excursion inside Theosophy has not been without challenges. The potential for social allotment, error of images and ceremonies, and a shallow mixing of customs have ignited reactions. The fragile harmony between conscious commitment and the gamble of weakening the realness of social articulations stays a continuous thought inside Theosophical circles.

In exploring these difficulties, Theosophy has tried to cultivate a genuine and deferential commitment with assorted social and otherworldly articulations. The accentuation on the all inclusiveness of profound standards, the acknowledgment of a typical wellspring of intelligence across customs, and the advancement of shared understanding have been fundamental to Theosophical endeavors to connect social partitions.

Chapter 3

Core Principles of Theorophy

Theorophy, a term got from the Greek words "theos" meaning God and "philo" significance love, is a complex philosophical structure that digs into the idea of the heavenly, the universe, and human life. At its center, Theorophy tries to orchestrate profound experiences with sane request, planning to overcome any issues between the magical and the coherent. In this investigation of Theorophy, we will dig into its center standards, analyzing the many-sided embroidery that winds around together powerful contemplations, moral reflections, and epistemological requests.

At the core of Theorophy lies a significant appreciation for the interconnectedness, everything being equal. Theorophical thought sets that the heavenly pervades each part of presence, from the loftiness of inestimable peculiarities to the details of daily existence. This interconnectedness fills in as the establishment for understanding the solidarity of the universe and the innate heavenly nature that joins all creatures.

In considering the idea of the heavenly inside Theorophy, one experiences a nuanced point of view that rises above unbending creeds and embraces a range of convictions. Theorophy recognizes the variety of strict and profound customs, remembering them as changed articulations of humankind's aggregate mission for importance and association with the heavenly. It looks to distil the consistent ideas woven through these assorted accounts, stressing the comprehensiveness of profound insights that resound across societies and ages.

Vital to Theorophy is the idea of greatness — an acknowledgment that the heavenly rises above human understanding and language. This affirmation of the unutterable idea of the heavenly calls for modesty notwithstanding the secrets that encompass presence. Theorophical scholars frequently advocate for a harmony among reason and instinct, empowering people to investigate the profundities of their internal being while likewise captivating in normal request to grow how they might interpret the heavenly.

Moral contemplations inside Theorophy spin around the possibility that the acknowledgment of heavenly interconnectedness intrinsically infers an obligation to develop sympathy and compassion. Theorophical morals accentuates the interconnectedness of every single living being and the ethical basic to contribute emphatically to the prosperity of the bigger inestimable embroidered artwork. This moral system reaches out past human-driven concerns, embracing a natural cognizance that perceives the relationship of all living things.

In the domain of epistemology, Theorophy explores the fragile harmony among confidence and reason. While recognizing the significance of exact request and consistent examination, Theorophical epistemology likewise perceives the constraints of simply materialistic viewpoints. It sets that specific parts of the real world, especially those connected with the heavenly and mystical domains, may evade observational confirmation and require a more broad epistemic methodology.

A pivotal part of Theorophy includes the investigation of the human mind and the groundbreaking capability of profound practices. Theorophical customs frequently consolidate thoughtful and reflective practices for of adjusting people to the heavenly inside and encouraging inward development. These practices expect to work with an immediate encounter of the otherworldly, rising above the limits of scholarly comprehension and welcoming people into a more profound, more significant association with the heavenly.

As Theorophy draws in with the perpetual inquiries of presence, it wrestles with the idea of misery and the quest for importance despite life's difficulties. Theorophical viewpoints on experiencing frequently draw different strict and philosophical practices, offering bits of knowledge into the extraordinary force of affliction and the potential for profound development despite life's innate battles.

Theorophy, as a philosophical structure, likewise addresses the lasting strain between individual independence and the aggregate prosperity. It advocates for the acknowledgment of individual organization while accentuating the significance of orchestrating individual yearnings with the more prominent astronomical request. This fragile harmony among independence and interconnectedness is viewed as fundamental for encouraging a general public that values both individual flexibility and aggregate congruity.

In the investigation of time and forever, Theorophy welcomes consideration on the idea of transience and the everlasting. Time, inside Theorophical thought, is in many cases seen as a unique embroidery where past, present, and future combine in a grandiose dance. The idea of forever stretches out past straight time, welcoming people to rise above the fleeting limitations of the material world and interface with the ageless pith of the heavenly.

A focal subject in Theorophy is arousing — the acknowledgment of one's heavenly nature and the acknowledgment of the interconnectedness, everything being equal. This enlivening isn't only a scholarly seeing however a significant change in cognizance that pervades each part of a singular's life. Theorophical customs frequently set that this enlivening is definitely not a one-time occasion yet a nonstop course of developing knowledge and change.

Theorophy likewise digs into the idea of images and models as vehicles for conveying otherworldly bits of insight. Images, inside Theorophical customs, are viewed as scaffolds between the commonplace and the heavenly, offering a method for communicating significant profound experiences that rise above semantic impediments.

Paradigms, established in the aggregate oblivious, act as widespread images that reverberate across societies and ages, reflecting key parts of the human experience.

Theorophical thought urges an integrative way to deal with information that draws from assorted sources, including strict texts, philosophical compositions, and logical disclosures. It looks to orchestrate apparently unique strands of intelligence, perceiving the correlative idea of various types of information. This integrative viewpoint cultivates an all encompassing comprehension of reality that rises above thin disciplinary limits.

Theorophy, in its investigation of the heavenly, likewise draws in with the idea of heavenly characteristic — the possibility that the heavenly isn't just otherworldly yet in addition natural inside the texture of creation. This characteristic welcomes people to perceive the holiness intrinsic in all parts of presence, cultivating a feeling of respect for the regular world and the interconnected snare of life.

Theorophical viewpoints on death and the hereafter fluctuate, incorporating a scope of convictions that mirror the variety of strict and social practices. Normal topics incorporate the thought of the spirit's excursion past actual passing, the potential for otherworldly development in the hereafter, and the possibility that demise is a progress as opposed to an endpoint. Theorophical thought frequently urges people to stand up to their mortality with a feeling of poise, seeing passing as a characteristic piece of the enormous request.

In the investigation of mystery inside Theorophy, the accentuation is put on immediate, experiential information on the heavenly. Otherworldly encounters are viewed for of rising above normal cognizance and experiencing the heavenly in a profoundly private and groundbreaking manner. Theorophical supernatural quality frequently includes rehearses that work with this immediate fellowship with the heavenly, like reflection, examination, and custom.

Theorophy likewise addresses the job of local area and the significance of otherworldly cooperation on the excursion of self-disclosure. Local area, inside Theorophical customs, is seen as a steady climate where people can share their experiences, participate in aggregate otherworldly practices, and cultivate a feeling of common perspective. This mutual aspect is viewed as essential to the most common way of enlivening and the development of an amicable society.

As Theorophy draws in with the difficulties of the contemporary world, it underlines the requirement for an all encompassing way to deal with worldwide issues. Ecological stewardship, civil rights, and moral administration are seen as interconnected features of a bigger vision for an amicable and supportable world. Theorophical points of view on cultural difficulties draw from moral standards and otherworldly bits of knowledge, upholding for a careful and merciful way to deal with resolving the complicated issues confronting humankind.

Theorophy, in its investigation of language and imagery, perceives the constraints of semantic articulation in conveying the unspeakable idea of the heavenly. Language, inside Theorophical thought, is seen as a representative instrument that focuses to more profound bits of insight instead of catching the completion of the heavenly reality. This etymological lowliness urges people to move toward consecrated texts and profound lessons with a nuanced understanding, perceiving the emblematic idea of strict language.

Theorophical points of view on instruction underline the significance of developing insight close by information. Training, inside Theorophical customs, is viewed as a groundbreaking cycle that stretches out past the securing of statistical data points. It incorporates the improvement of moral qualities, decisive reasoning abilities, and an extending consciousness of the interconnectedness, everything being equal. Theorophical training looks to support people who are mentally capable as well as ethically and profoundly cognizant.

In the domain of feel, Theorophy investigates the significant association among magnificence and the heavenly. The enthusiasm for excellence, whether communicated through craftsmanship, music, or nature, is seen for of adjusting the spirit to the extraordinary. Theorophical style urges people to perceive the heavenly brilliance reflected in the excellence of the world, encouraging a feeling of stunningness and respect for the imaginative rule that pervades the universe.

3.1 Identifying and explicating the fundamental principles that underlie Theorophy.

Distinguishing and elucidating the key rules that underlie Theorophy requires a nuanced investigation of its center fundamentals, which incorporate otherworldly, moral, epistemological, and functional aspects. At its quintessence, Theorophy addresses a philosophical structure that tries to incorporate profound bits of knowledge with judicious request, encouraging a comprehensive comprehension of presence. This assessment will dig into the primary rules that structure the foundation of Theorophy, revealing insight into its viewpoints on the heavenly, interconnectedness, morals, information, and the extraordinary capability of human cognizance.

The supernatural component of Theorophy rotates around the major idea of the heavenly and its relationship to the universe. Vital to Theorophical believed is the acknowledgment that the heavenly, frequently alluded to as God or the Outright, is the basic pith that pervades all of presence. This heavenly the truth is viewed as otherworldly, outperforming human understanding and language, yet natural inside each part of creation. Theorophy places that the universe is definitely not a disconnected assortment of unique components yet an amicable and interconnected entire, with the heavenly filling in as the binding together rule that ties everything together.

In investigating the idea of the heavenly, Theorophy takes on a pluralistic viewpoint that embraces the variety of strict and otherworldly practices. As opposed to demanding a solitary, narrow minded understanding, Theorophical thought perceives the legitimacy of different strict stories and practices as different articulations of mankind's mission for significance and association with the heavenly. This comprehensive position cultivates an appreciation for the all inclusiveness of otherworldly bits of insight, rising above social and verifiable limits.

A critical idea inside Theorophy is greatness, which underlines the indescribable and extraordinary nature of the heavenly. Theorophical scholars declare that the heavenly can't be completely caught or epitomized by human language or reasonable structures. This affirmation of the constraints of language and mind requires an unassuming acknowledgment of the secrets that encompass the heavenly, empowering people to approach the consecrated with stunningness, worship, and a feeling of miracle.

Theorophical mysticism likewise wrestles with the idea of interconnectedness, stating that each component of the universe is unpredictably connected in a tremendous and entwined embroidery. This interconnectedness reaches out past the material domain, including the profound and magical aspects. The acknowledgment of interconnectedness fills in as the reason for moral contemplations inside Theorophy, underscoring the ethical basic to develop sympathy, compassion, and a feeling of obligation toward every living being.

Morals inside Theorophy is established in the comprehension that the interconnectedness of everything suggests a common obligation regarding the prosperity of the bigger grandiose request. Theorophical morals rises above anthropocentrism, perceiving the natural worth of all living things and the interconnected snare of presence. This moral system reaches out past traditional moral contemplations and embraces a biological cognizance, upholding for the dependable stewardship of the climate and an agreeable concurrence with the normal world.

Theorophy, in its moral reflections, accentuates the significance of adjusting individual activities to the more prominent astronomical congruity. Theorophical morals urges people to develop ideals like empathy, lowliness, and uprightness, remembering them as articulations of the heavenly inside human awareness. This moral direction isn't exclusively worried about private excellence yet stretches out to the cultural and worldwide levels, advancing civil rights, value, and a feeling of obligation toward people in the future.

In the domain of epistemology, Theorophy explores the mind boggling connection among confidence and reason. While recognizing the worth of experimental request and levelheaded examination, Theorophical epistemology places that specific parts of the real world, especially those connected with the heavenly and powerful domains, may escape observational check. This acknowledgment requires a more extensive epistemic methodology that envelops both explanation and instinct, welcoming people to investigate the profundities of their internal being and participate in thoughtful practices to get to higher types of information.

Theorophical epistemology likewise recognizes the constraints of language in conveying otherworldly bits of insight. Language, inside Theorophical thought, is viewed as an emblematic instrument that focuses toward more profound real factors as opposed to exemplifying them completely. This semantic modesty empowers a nuanced and scrutinizing way to deal with holy texts, perceiving the emblematic idea of strict language and the requirement for translation that goes past peculiarity.

The groundbreaking capability of human cognizance is a focal subject in Theorophy. Theorophical customs frequently integrate pondering and reflective practices for the purpose of adjusting people to the heavenly inside and working with internal development. These practices are not just elusive ceremonies but rather commonsense roads for people to encounter direct fellowship with the otherworldly, rising above scholarly comprehension and encouraging a profound and special interaction with the heavenly.

Theorophical thought resolves the enduring inquiries of presence, including the idea of misery and the quest for importance despite life's difficulties. Theorophical viewpoints on experiencing draw different strict and philosophical practices, offering bits of knowledge into the extraordinary force of difficulty and the potential for otherworldly development notwithstanding life's inborn battles. Theorophy urges people to move toward enduring with versatility and a feeling of direction, seeing it as an impetus for individual and profound turn of events.

The harmony between individual independence and the aggregate prosperity is an essential thought inside Theorophy. It perceives the significance of individual organization while accentuating the requirement for blending individual desires with the more noteworthy inestimable request. The sensitive harmony among independence and interconnectedness is viewed as fundamental for cultivating a general public that values both individual opportunity and aggregate concordance. Theorophical points of view on friendly elements advocate for moral administration, civil rights, and a guarantee to making networks that mirror the standards of interconnectedness and sympathy.

In the investigation of time and endlessness, Theorophy welcomes consideration on the idea of fleetingness and the everlasting. Time, inside Theorophical thought, is in many cases seen as a powerful embroidery where past, present, and future blend in a grandiose dance. The idea of forever stretches out past direct time, welcoming people to rise above the transient limitations of the material world and interface with the ageless substance of the heavenly. This transient viewpoint impacts Theorophical sees on the repeating idea of presence, rebirth, and the possibility that the spirit's process stretches out past a solitary lifetime.

A focal topic in Theorophy is the idea of enlivening — an experiential acknowledgment of one's heavenly nature and the interconnectedness, everything being equal. This enlivening is definitely not a simple scholarly seeing yet a significant change in cognizance that penetrates each part of a singular's life.

Theorophical customs frequently place that this enlivening is a consistent interaction, welcoming people into a more profound, more significant association with the heavenly and an elevated consciousness of the reliance of all life.

Images and models assume a critical part in Theorophical thought as vehicles for conveying otherworldly insights. Images are seen as scaffolds between the ordinary and the heavenly, offering a method for communicating significant otherworldly experiences that rise above etymological constraints. Models, established in the aggregate oblivious, act as general images that reverberate across societies and ages, reflecting key parts of the human experience. Theorophical viewpoints on images and paradigms stress their extraordinary expected in directing people toward a more profound comprehension of the heavenly and the secrets of presence.

Theorophical thought urges an integrative way to deal with information that draws from different sources, including strict texts, philosophical compositions, and logical revelations. This integrative viewpoint encourages a comprehensive comprehension of reality that rises above limited disciplinary limits. Theorophy welcomes people to perceive the corresponding idea of various types of information, appreciating the lavishness that arises when different bits of knowledge are woven together into an intelligible embroidery of understanding.

Theorophy, in its investigation of language and imagery, perceives the constraints of semantic articulation in conveying the unspeakable idea of the heavenly. Language, inside Theorophical thought, is viewed as an emblematic device that focuses to more profound insights instead of catching the completion of the heavenly reality. This semantic lowliness urges people to move toward consecrated texts and otherworldly lessons with a nuanced understanding, perceiving the emblematic idea of strict language and the requirement for interpretive profundity.

Theorophical points of view on death and eternity envelop a scope of convictions that mirror the variety of strict and social customs. Normal subjects incorporate the thought of the spirit's excursion past actual passing, the potential for otherworldly development in eternity, and the possibility that demise is a progress as opposed to an endpoint. Theorophical thought frequently urges people to stand up to their mortality with composure, seeing demise as a characteristic piece of the infinite request and a chance for the spirit to proceed with its excursion.

Mystery holds a huge spot inside Theorophy, underscoring immediate, experiential information on the heavenly. Otherworldly encounters are viewed as a method for rising above conventional cognizance and experiencing the heavenly in a profoundly private and extraordinary manner. Theorophical mystery frequently includes practices like reflection, thought, and custom, filling in as pathways for people to extend their association with the extraordinary and experience a significant feeling of solidarity with the heavenly.

Local area and the significance of otherworldly partnership are necessary parts of Theorophical thought. Local area is seen as a steady climate where people can share their experiences, participate in aggregate otherworldly practices, and encourage a feeling of common perspective. This public aspect is viewed as fundamental for the most common way of enlivening and the development of an amicable society. Theorophy energizes the development of networks that exemplify the standards of interconnectedness, sympathy, and common help.

As Theorophy draws in with the difficulties of the contemporary world, it stresses the requirement for a comprehensive way to deal with worldwide issues. Ecological stewardship, civil rights, and moral administration are seen as interconnected features of a bigger vision for an agreeable and feasible world. Theorophical points of view on cultural difficulties draw from moral standards and otherworldly experiences, pushing for a careful and merciful way to deal with resolving the mind boggling issues confronting humankind.

Theorophy, as a living way of thinking, adjusts and develops in light of the changing elements of human culture and the continuous investigation of the secrets of presence. It's anything but an unbending creed however a unique system that welcomes people to take part in a persistent course of request, reflection, and individual change. Theorophical thought energizes a receptive and comprehensive methodology, embracing the lavishness of human variety while perceiving the fundamental solidarity that ties all of creation.

All in all, Theorophy remains as a significant and complex philosophical system that resolves the central inquiries of human life. Its center standards, established in the acknowledgment of the heavenly, interconnectedness, moral obligation, and the groundbreaking capability of human cognizance, offer a comprehensive point of view on the idea of the real world. Theorophy welcomes people into an excursion of self-disclosure, encouraging a developing consciousness of the heavenly inside and motivating a pledge to contribute decidedly to the more prominent infinite embroidery of life. As we dive into the fundamental standards of Theorophy, we experience a rich and enlightening investigation of the secrets of the universe and the immortal mission for significance and association.

3.2 Exploring the interconnectedness of spirituality, philosophy, and mysticism within Theorophical teachings.

Investigating the interconnectedness of otherworldliness, reasoning, and magic inside Theorophical lessons uncovers a rich embroidery of imagined that rises above traditional limits and welcomes people to leave on a significant excursion of self-disclosure and profound investigation.

At the center of Theorophical lessons is the acknowledgment of otherworldliness as a natural part of the human experience. Otherworldliness, inside Theorophy, isn't restricted to strict creeds or regulated rehearses yet is perceived as a widespread component of human cognizance that looks for association with the heavenly.

This sweeping perspective on otherworldliness envelops a range of convictions, encounters, and practices that mirror the variety of human profound articulation.

Reasoning, as an ally to otherworldliness inside Theorophical lessons, fills in for of scholarly request and basic reflection. Theorophical reasoning tries to orchestrate profound bits of knowledge with objective investigation, overcoming any barrier between the magical and the sensible. It gives a structure to grasping the idea of the heavenly, the universe, and human life, offering a sound and integrative point of view that rises above the impediments of limited disciplinary limits.

The investigation of enchantment inside Theorophy develops the association between the individual and the heavenly. Enchantment, in Theorophical terms, alludes to the immediate, experiential information on the heavenly that rises above common awareness. An extraordinary excursion includes rising above the constraints of the inner self and getting to significant conditions of solidarity with the heavenly. Supernatural encounters, frequently worked with through scrutinizing practices like reflection and petition, are seen for the purpose of adjusting the person to the otherworldly components of the real world.

The interconnectedness of otherworldliness, reasoning, and magic inside Theorophical lessons is clear in the manner these aspects illuminate and advance one another. Otherworldliness gives the establishment, reasoning offers scholarly intelligence, and magic works with direct fellowship with the heavenly. Together, they make a comprehensive structure that resolves the major inquiries of presence and urges people to explore the profundities of their internal being.

Inside Theorophical otherworldliness, the heavenly is seen as the hidden embodiment that penetrates all of presence. This heavenly the truth is both otherworldly and inherent, rising above human cognizance while being available inside each part of creation. Theorophical otherworldliness stresses an individual and direct association with the heavenly, rising above the bounds of standardized strict designs. It urges people to develop a profound and close connection with the heavenly, encouraging a feeling of wonderment, veneration, and love for the hallowed.

Theorophical otherworldliness likewise embraces a pluralistic viewpoint that recognizes the legitimacy of different strict and profound customs. As opposed to advancing selectiveness or unyieldingness, Theorophical lessons perceive the comprehensiveness of profound insights that manifest in different social and authentic settings. This comprehensive methodology urges people to appreciate and gain from the insight implanted in various strict stories, cultivating a feeling of solidarity that rises above partisan partitions.

Reasoning, inside Theorophical lessons, fills in as a scaffold between the natural experiences of otherworldliness and the scientific meticulousness of scholarly request.

Theorophical reasoning explores the intricacies of power, epistemology, and morals, giving a rational system to figuring out the idea of the real world and the human experience. It urges people to take part in basic reflection, to address presumptions, and to investigate the more profound ramifications of otherworldly bits of knowledge inside a normal setting.

Mysticism, as a part of Theorophical reasoning, digs into the idea of the heavenly and the basic design of the universe. Theorophical power sets that the universe is an amicable and interconnected entire, with the heavenly filling in as the bringing together rule that ties everything together. This otherworldly point of view illuminates

Theorophical otherworldliness, giving a reasonable establishment to understanding the heavenly as both extraordinary and inborn.

Epistemology, inside Theorophical reasoning, resolves questions connected with information and the manners by which people come to figure out the heavenly. Theorophical epistemology perceives the restrictions of simply experimental methodologies, particularly while investigating parts of reality connected with the heavenly and the supernatural. It advocates for a more far reaching epistemic methodology that incorporates natural bits of knowledge, scrutinizing encounters, and the acknowledgment of the constraints of language in catching otherworldly bits of insight.

Morals, as one more element of Theorophical reasoning, originates from the interconnectedness, everything being equal. Theorophical morals underscores the ethical basic to develop sympathy, compassion, and a feeling of obligation toward every living being. This moral structure reaches out past human-driven worries to incorporate natural cognizance and a pledge to dependable stewardship of the climate. Theorophical morals, established in the acknowledgment of heavenly interconnectedness, gives a directing compass to people trying to adjust their activities to the more noteworthy vast congruity.

Supernatural quality, as a necessary part of Theorophical lessons, gives an immediate and experiential aspect to the interconnected snare of otherworldliness and reasoning. Magic includes rising above normal awareness to straightforwardly experience the heavenly. Theorophical otherworldliness frequently utilizes scrutinizing rehearses, like reflection and petition, to work with these groundbreaking encounters. Mysterious experiences are seen for of rising above the self image, getting to higher conditions of cognizance, and developing one's association with the heavenly.

The interconnectedness of otherworldliness, reasoning, and magic inside Theorophical lessons is additionally exemplified by the accentuation on the groundbreaking capability of human awareness. Theorophy sees the excursion of self-disclosure and profound arousing as key to the human experience.

Thoughtful practices, directed by otherworldly experiences, are viewed as pragmatic roads for people to stir to the heavenly inside and cultivate internal development. Theorophical lessons empower a constant course of extending understanding and change, welcoming people to take part in rehearses that develop an uplifted familiarity with the interconnectedness of all life.

Theorophical points of view on affliction and the quest for importance notwithstanding life's difficulties draw from the insight of otherworldliness, reasoning, and magic. The acknowledgment of enduring as an inborn piece of the human condition is met with the comprehension that difficulty can be an impetus for otherworldly development. Theorophy urges people to stand up to life's battles with versatility and reason, seeing difficulties as any open doors for self-revelation and the refinement of character.

The harmony between individual independence and the aggregate prosperity is a repetitive subject inside the interconnected texture of Theorophy. Theorophical lessons perceive the significance of individual organization while accentuating the need to blend individual yearnings with the more prominent inestimable request. This sensitive equilibrium is viewed as fundamental for encouraging a general public that values both individual flexibility and aggregate congruity. Theorophical viewpoints on friendly elements advocate for moral administration, civil rights, and a pledge to making networks that mirror the standards of interconnectedness and sympathy.

In the investigation of time and forever, Theorophy welcomes examination on the idea of fleetingness and the everlasting. Time, inside Theorophical thought, is in many cases seen as a unique embroidery where past, present, and future combine in a vast dance. The idea of forever reaches out past direct time, welcoming people to rise above the transient requirements of the material world and associate with the ageless substance of the heavenly. This transient viewpoint impacts Theorophical sees on the repeating idea of presence, resurrection, and the possibility that the spirit's process reaches out past a solitary lifetime.

Images and originals assume a critical part in conveying extraordinary insights inside Theorophical thought. Images are viewed as extensions between the unremarkable and the heavenly, offering a method for communicating significant otherworldly bits of knowledge that rise above phonetic restrictions. Originals, established in the aggregate oblivious, act as general images that reverberate across societies and ages, reflecting principal parts of the human experience. Theorophical viewpoints on images and paradigms underscore their groundbreaking expected in directing people toward a more profound comprehension of the heavenly and the secrets of presence.

Theorophical thought urges an integrative way to deal with information that draws from different sources, including strict texts, philosophical compositions, and logical disclosures. This integrative point of view cultivates an all encompassing comprehension of reality that rises above slender disciplinary limits.

Theorophy welcomes people to perceive the correlative idea of various types of information, appreciating the lavishness that arises when different experiences are woven together into a cognizant embroidery of understanding.

Theorophy, in its investigation of language and imagery, perceives the constraints of etymological articulation in conveying the unspeakable idea of the heavenly. Language, inside Theorophical thought, is viewed as an emblematic device that focuses to more profound bits of insight instead of catching the completion of the heavenly reality. This semantic modesty urges people to move toward sacrosanct texts and otherworldly lessons with a nuanced understanding, perceiving the emblematic idea of strict language and the requirement for interpretive profundity.

Demise and eternity, inside Theorophical lessons, envelop a scope of convictions that mirror the interconnectedness of otherworldliness, reasoning, and supernatural quality. Theorophical points of view on death frequently accentuate the thought of the spirit's excursion past actual mortality. Theorophical thought sees passing as a progress as opposed to an endpoint, seeing it as a characteristic piece of the inestimable request and a chance for the spirit to proceed with its otherworldly development.

The interconnectedness of life and demise, as investigated inside Theorophical lessons, highlights the repetitive idea of presence and the everlasting part of the spirit's excursion.

Mystery holds a focal spot inside the interconnected texture of otherworldliness, reasoning, and magic inside Theorophy. Otherworldly encounters are viewed as an immediate and prompt experience with the heavenly, rising above the restrictions of calculated understanding. Theorophical otherworldliness frequently includes practices like reflection, consideration, and petitioning heaven as method for working with these groundbreaking encounters. Mystery fills in as a binding together string, meshing together otherworldliness and theory into a lived and experiential element of Theorophical lessons.

Local area, inside Theorophical thought, is seen as a vital part of the profound excursion and an indication of the interconnectedness, everything being equal. Otherworldly partnership gives people a steady climate where they can share bits of knowledge, participate in aggregate practices, and encourage a feeling of common perspective. Theorophical lessons underscore the significance of profound local area as a space for common help, learning, and the development of an agreeable and caring lifestyle.

As Theorophy draws in with the difficulties of the contemporary world, it underscores the requirement for a comprehensive way to deal with worldwide issues. Natural stewardship, civil rights, and moral administration are seen as interconnected features of a bigger vision for an agreeable and practical world. Theorophical viewpoints on cultural difficulties draw from moral standards and profound experiences, supporting for a careful and merciful way to deal with resolving the perplexing issues confronting mankind.

3.3 Illustrating how these principles provide a holistic understanding of existence.

Delineating how the standards of Theorophy give an all encompassing comprehension of presence includes diving into the complex embroidery woven by its magical, moral, epistemological, and useful aspects. These standards, profoundly interconnected, make a far reaching system that rises above reductionist viewpoints and welcomes people to investigate the significant interchange between the material and otherworldly domains.

At the center of Theorophical standards is the supernatural comprehension of the heavenly as the hidden embodiment penetrating all of presence. This heavenly reality, both extraordinary and inborn, fills in as the binding together rule that ties the universe together. Theorophy places that the universe is definitely not an incoherent assortment of separated components yet an agreeable entire, complicatedly interconnected and suffused with the hallowed. This mystical point of view establishes the groundwork for a comprehensive comprehension of presence that stretches out past the limits of realism and reductionism.

The acknowledgment of interconnectedness is a focal precept of Theorophical mysticism, enlightening the complex snare of connections that characterize the universe. The interconnectedness reaches out past the actual domain to envelop the profound and supernatural aspects, uncovering a consistent solidarity that rises above dualistic reasoning. This standard highlights the reliance of every living being and the environmental awareness that perceives the inborn worth of each and every component inside the inestimable woven artwork.

Morals inside Theorophy arises naturally from the supernatural standards of interconnectedness and heavenly solidarity. The acknowledgment that all creatures share a typical otherworldly substance leads to a moral system that rises above anthropocentrism. Theorophical morals stresses sympathy, compassion, and a feeling of obligation toward every living being. This moral direction reaches out past individual activities to address more extensive cultural and ecological worries, mirroring an all encompassing way to deal with moral living.

Theorophical epistemology, the part of theory worried about the nature and cutoff points of information, supplements the mystical and moral aspects by exploring the fragile harmony among confidence and reason. Theorophy recognizes the worth of experimental request and judicious examination while perceiving the impediments of absolutely materialistic points of view. This epistemological position energizes a receptive investigation of instinctive bits of knowledge, thoughtful encounters, and the acknowledgment of the unspeakable idea of the heavenly.

The extraordinary capability of human cognizance, a foundation of Theorophy, adds a down to earth and experiential aspect to its all encompassing comprehension of presence. Theorophical lessons underline scrutinizing practices like contemplation, petition, and thoughtfulness as method for adjusting people to the heavenly inside.

These practices go outside scholarly ability to grasp, offering an immediate and groundbreaking experience of the interconnectedness, everything being equal. Theorophy considers individual change to be necessary to the more extensive development of human cognizance and the acknowledgment of an amicable presence.

In delineating how these standards give a comprehensive comprehension of presence, it is fundamental to investigate the viable ramifications of Theorophical thought in resolving the perpetual inquiries of human experience. The mystical rule of interconnectedness, for instance, illuminates Theorophical points of view on anguish. The acknowledgment that all creatures are interconnected cultivates a caring reaction to the battles of others, underlining compassion and a common obligation regarding easing experiencing on the planet.

Theorophical sees on time and forever add to its all encompassing comprehension of presence by welcoming people to rise above regular thoughts of fleetingness. Theorophy places that time isn't simply a straight movement yet a powerful embroidery where past, present, and future blend. The idea of endlessness, in Theorophical terms, rises above the impediments of straight time, welcoming people to associate with the ageless pith of the heavenly. This transient viewpoint shapes Theorophical viewpoints on the repeating idea of presence and the timeless part of the spirit's excursion.

Images and paradigms, major to Theorophical thought, assume a significant part in giving an all encompassing comprehension of presence. Images are viewed as extensions between the commonplace and the heavenly, offering a language that rises above the limits of normal talk. Models, established in the aggregate oblivious, act as widespread images that resound across societies and ages, reflecting immortal insights about the human experience. Theorophical points of view on images and models stress their groundbreaking possible in directing people toward a more profound comprehension of the secrets of presence.

The coordination of assorted wellsprings of information, one more standard of Theorophy, adds to its comprehensive methodology. Theorophical thought urges people to draw from strict texts, philosophical compositions, and logical disclosures, perceiving the correlative idea of various types of information. This integrative viewpoint encourages a comprehensive comprehension of reality that rises above limited disciplinary limits, welcoming people to see the value in the lavishness that arises when different experiences are woven together into a lucid embroidery of understanding.

Theorophical otherworldliness, as a major part of its all encompassing perspective, supports an individual and direct association with the heavenly. The acknowledgment of the heavenly as both extraordinary and characteristic shapes Theorophical sees on the holiness inborn in all parts of presence.

Theorophy considers otherworldliness to be a general component of human cognizance, rising above strict doctrines and systematized rehearses. This comprehensive methodology highlights the all inclusiveness of profound bits of insight that manifest in different social and verifiable settings.

Reasoning inside Theorophy fills in as an extension between the natural experiences of otherworldliness and the scientific meticulousness of scholarly request. Theorophical reasoning explores magical inquiries, moral contemplations, and epistemological difficulties, giving a sound structure to figuring out the idea of the real world. Theorophy urges people to participate in basic reflection, to address presumptions, and to investigate the more profound ramifications of otherworldly experiences inside a reasonable setting.

Mystery, an essential part of Theorophical lessons, gives an immediate and experiential aspect to its all encompassing comprehension of presence. Otherworldliness includes rising above customary cognizance to straightforwardly experience the heavenly. Theorophical enchantment frequently utilizes thoughtful practices, like reflection and supplication, as method for working with these extraordinary encounters. Magic fills in as a binding together string, meshing together otherworldliness and theory into a lived and experiential element of Theorophical lessons.

Theorophy's accentuation on private change lines up with its comprehensive way to deal with grasping presence. The groundbreaking excursion, worked with through pensive practices, includes rising above the constraints of the self image and getting to higher conditions of cognizance. Theorophy considers self-awareness and profound arousing to be fundamental to the more extensive development of human cognizance and the acknowledgment of an amicable presence.

In showing how these standards give an all encompassing comprehension of presence, it is fundamental to perceive their interconnected nature and the synergistic impacts they produce. The magical acknowledgment of the heavenly as the fundamental quintessence illuminates moral contemplations, forming a merciful and environmentally cognizant way to deal with living. Theorophical epistemology explores the sensitive harmony among confidence and reason, empowering a nuanced investigation of information that rises above unbending polarities.

The groundbreaking capability of human cognizance, established in Theorophy's magical and moral standards, unfurls as a lived insight through pensive practices. The interconnectedness, everything being equal, a core value, impacts Theorophical points of view on misery, time, and endlessness. Images and paradigms, drawn from the aggregate oblivious, add to an emblematic language that rises above semantic limits, directing people toward a more profound comprehension of the secrets of presence.

The integrative way to deal with information inside Theorophy mirrors its obligation to a comprehensive comprehension of the real world. By drawing from different sources, Theorophy empowers a rich and diverse investigation of presence that rises above thin disciplinary limits. Theorophical otherworldliness, reasoning, and magic combine to make an exhaustive system that resolves the crucial inquiries of human experience, giving a guide to people looking for a more profound comprehension of the interconnectedness of all life.

As Theorophy draws in with the difficulties of the contemporary world, it highlights the requirement for a comprehensive way to deal with worldwide issues. Ecological stewardship, civil rights, and moral administration are seen as interconnected features of a bigger vision for an amicable and practical world. Theorophical points of view on cultural difficulties draw from moral standards and otherworldly experiences, upholding for a careful and merciful way to deal with resolving the complicated issues confronting mankind.

A comprehensive comprehension of presence inside the system of Theorophy includes a multi-layered investigation of mystical, moral, epistemological, and reasonable aspects. The interconnected standards of Theorophy make an extensive perspective that rises above reductionist points of view, offering people a significant focal point through which to understand the complexities of life, cognizance, and the universe.

At the powerful center of Theorophy lies the acknowledgment of the heavenly as the hidden quintessence saturating all of presence. This heavenly the truth is described as both extraordinary, outperforming human understanding, and characteristic, existing inside each feature of creation. Theorophy places that the universe is certainly not a disconnected assortment of different components yet an amicable entire, unpredictably interconnected and suffused with the consecrated. This powerful viewpoint establishes the groundwork for a comprehensive comprehension of presence that stretches out past realism and reductionism, welcoming people to consider the significant interconnectedness, everything being equal.

Interconnectedness, a key standard of Theorophy, fills in as a directing string that winds through the texture of presence. This guideline reaches out past the actual domain to incorporate the otherworldly and powerful aspects, uncovering a consistent solidarity that rises above dualistic reasoning. The interconnectedness of every single living being and the natural awareness that emerges from this acknowledgment structure the reason for a moral system inside Theorophy. It calls for sympathy, compassion, and a feeling of obligation toward all life, encouraging a comprehensive way to deal with moral living that rises above tight anthropocentrism.

Morals inside Theorophy arises naturally from the magical standards of interconnectedness and heavenly solidarity. The acknowledgment that all creatures share a typical profound pith leads to a moral system that rises above individual activities and addresses more extensive cultural and ecological worries.

Theorophical morals stresses empathy, natural stewardship, and a guarantee to capable living that lines up with the more noteworthy infinite congruity. This moral direction adds to the comprehensive comprehension of presence by perceiving the relationship of moral decisions and the prosperity of the whole interconnected trap of life.

Theorophical epistemology, worried about the nature and cutoff points of information, supplements the otherworldly and moral aspects by exploring the fragile harmony among confidence and reason. Theorophy recognizes the worth of exact request and reasonable investigation while perceiving the impediments of absolutely materialistic viewpoints. This epistemological position empowers a receptive investigation of instinctive bits of knowledge, scrutinizing encounters, and the acknowledgment of the unspeakable idea of the heavenly. Theorophical epistemology welcomes people to embrace a more extensive comprehension of information that includes both reasonable and natural methods of dread, adding to the all encompassing perspective of Theorophy.

The groundbreaking capability of human cognizance, a foundation of Theorophy, adds a down to earth and experiential aspect to its comprehensive comprehension of presence. Theorophical lessons stress insightful practices like reflection, petition, and contemplation as method for adjusting people to the heavenly inside. These practices go outside scholarly ability to comprehend, offering an immediate and extraordinary experience of the interconnectedness, everything being equal. Theorophy considers individual change to be essential to the more extensive development of human cognizance and the acknowledgment of an amicable presence.

In representing how these standards give a comprehensive comprehension of presence, it is fundamental to investigate the reasonable ramifications of Theorophical thought in resolving the perpetual inquiries of human experience. The powerful rule of interconnectedness illuminates Theorophical viewpoints on torment. The acknowledgment that all creatures are interconnected cultivates an empathetic reaction to the battles of others, underlining compassion and a common obligation regarding reducing experiencing on the planet. Theorophy sees experiencing not as disconnected occasions but rather as interconnected strings in the texture of presence, requiring a merciful and comprehensive reaction.

Theorophical sees on time and forever add to its all encompassing comprehension of presence by welcoming people to rise above traditional ideas of transience. Theorophy sets that time isn't only a direct movement however a powerful embroidery where past, present, and future blend. The idea of forever, in Theorophical terms, rises above the impediments of straight time, welcoming people to associate with the immortal quintessence of the heavenly. This fleeting viewpoint shapes Theorophical sees on the recurrent idea of presence and the timeless part of the spirit's excursion, offering an all encompassing viewpoint that rises above limited worldly requirements.

Images and paradigms, principal to Theorophical thought, assume an essential part in giving an all encompassing comprehension of presence. Images are viewed as scaffolds between the commonplace and the heavenly, offering a language that rises above the constraints of standard talk. Prime examples, established in the aggregate oblivious, act as widespread images that resound across societies and ages, reflecting immortal insights about the human experience. Theorophical viewpoints on images and models stress their groundbreaking likely in directing people toward a more profound comprehension of the secrets of presence. These emblematic portrayals offer a language that rises above phonetic constraints, giving a comprehensive and all inclusive method for investigating the significant elements of the real world.

The combination of different wellsprings of information, one more guideline of Theorophy, adds to its all encompassing methodology. Theorophical thought urges people to draw from strict texts, philosophical compositions, and logical revelations, perceiving the correlative idea of various types of information. This integrative point of view cultivates a comprehensive comprehension of reality that rises above limited disciplinary limits. Theorophy welcomes people to see the value in the wealth that arises when different bits of knowledge are woven together into a rational embroidery of understanding. Along these lines, Theorophy underscores the interconnectedness of information and the solidarity that underlies different articulations of shrewdness.

Theorophical otherworldliness, as a central part of its comprehensive perspective, empowers an individual and direct association with the heavenly. The acknowledgment of the heavenly as both extraordinary and natural shapes Theorophical sees on the holiness innate in all parts of presence. Theorophy considers otherworldliness to be a widespread element of human cognizance, rising above strict creeds and standardized rehearses. This comprehensive methodology highlights the comprehensiveness of otherworldly insights that manifest in different social and verifiable settings. Theorophical otherworldliness adds to the all encompassing comprehension of presence by stressing the interconnectedness of all life through a common profound pith.

Reasoning inside Theorophy fills in as a scaffold between the natural bits of knowledge of otherworldliness and the logical meticulousness of scholarly request.

Theorophical reasoning explores otherworldly inquiries, moral contemplations, and epistemological difficulties, giving an intelligible structure to figuring out the idea of the real world. Theorophy urges people to take part in basic reflection, to address presumptions, and to investigate the more profound ramifications of otherworldly bits of knowledge inside a sane setting. Theorophical reasoning adds to the all encompassing comprehension of presence by offering a scaffold between the natural and insightful components of human cognizance, encouraging a complete and integrative way to deal with information.

Mystery, an indispensable part of Theorophical lessons, gives an immediate and experiential aspect to its comprehensive comprehension of presence. Supernatural quality includes rising above normal awareness to straightforwardly experience the heavenly. Theorophical otherworldliness frequently utilizes insightful practices, like reflection and petition, as method for working with these extraordinary encounters. Supernatural quality fills in as a binding together string, meshing together otherworldliness and reasoning into a lived and experiential element of Theorophical lessons. Mysterious encounters add to the all encompassing comprehension of presence by furnishing people with direct admittance to the extraordinary components of the real world, encouraging an extended consciousness of the interconnectedness of all life.

Theorophy's accentuation on private change lines up with its all encompassing way to deal with figuring out presence. The extraordinary excursion, worked with through thoughtful practices, includes rising above the restrictions of the inner self and getting to higher conditions of awareness. Theorophy considers self-improvement and otherworldly arousing to be basic to the more extensive development of human cognizance and the acknowledgment of an amicable presence. This accentuation on private change adds to the comprehensive comprehension of presence by perceiving the interconnectedness of individual and aggregate advancement, underlining the job of self-awareness in the more extensive setting of enormous unfurling.

In representing how these standards give a comprehensive comprehension of presence, it is fundamental to perceive their interconnected nature and the synergistic impacts they produce. The otherworldly acknowledgment of the heavenly as the basic substance illuminates moral contemplations, molding a merciful and naturally cognizant way to deal with living. Theorophical epistemology explores the sensitive harmony between confidence.

Chapter 4

The Inner Journey

The inward excursion is a multifaceted investigation of oneself, a significant odyssey that rises above the outside world and dives profound into the openings of one's cognizance. A journey frequently stays concealed, concealed underneath the outer layer of day to day existence, yet its effect reverberates in each thought, feeling, and activity. The inward excursion is a journey of the spirit, a stay into the domain of self-disclosure and self-acknowledgment.

At the core of the inward excursion lies the mission for understanding, a tenacious pursuit to disentangle the secrets that exist in. An excursion starts with contemplation, a cognizant going internal to look at the layers of one's being. This thoughtful look is likened to stripping back the layers of an onion, uncovering the center substance that characterizes a person.

In the rushing about of current life, the inward excursion frequently assumes a lower priority in relation to the requests of the outside world. The chaos of day to day obligations, social assumptions, and outer tensions can overwhelm the inconspicuous murmurs of the spirit. However, the inward excursion calmly anticipates, prepared to unfurl when one decides to tune in.

The impetus for the internal excursion can change from one individual to another. For some's purposes, it could be a snapshot of emergency — an individual misfortune, a significant disillusionment, or an unexpected arousing that shakes the underpinnings of their reality. For other people, it could be a slow acknowledgment, a peaceful acknowledgment that there is something else to life besides what might be immediately obvious. No matter what the trigger, the inward excursion calls, welcoming people to leave on a way of self-investigation.

As one endeavors into the internal domains, the territory is both recognizable and unfamiliar. The scene is formed by recollections, encounters, and the engravings of the past. It is a territory set apart by the scars of wounds, the reverberations of giggling, and the shadows of unfulfilled dreams. Exploring this interior scene requires boldness, for it requests a fair a showdown with one's feelings of trepidation, frailties, and weaknesses.

The inward excursion is certainly not a straight movement yet a powerful cycle, a constant ebb, and stream of self-revelation. It includes going up against the shadows that sneak toward the sides of the psyche — those parts of the self that are frequently denied or smothered. Embracing the shadows is a fundamental stage in the internal excursion, for it is in recognizing the hazier parts of oneself that genuine change can happen.

At the center of the inward excursion is the investigation of personality. Who am I? What characterizes me? These inquiries reverberation through the offices of the spirit, inciting a mission for credibility. The veils worn to adjust to cultural assumptions are stripped away, uncovering the crude, unfiltered self underneath. It is a course of recovering one's actual substance and lining up with the inward compass that directs the excursion.

The inward excursion is likewise a journey for significance and reason. In a world that frequently appears to be turbulent and flighty, the quest for importance turns into a directing power. What is the reason for one's presence? What gives life meaning? These existential requests drive people to look for more profound associations — with themselves, with others, and with the universe in general.

All through the inward excursion, the idea of care turns into an important friend. Care is the act of being completely present at the time, developing an elevated familiarity with considerations, sentiments, and sensations. An instrument enables people to notice the functions of their psyches without judgment, cultivating a profound comprehension of oneself.

As the inward excursion unfurls, people might experience snapshots of isolation and quietness. In these thoughtful spaces, the commotion of the outside world disperses, considering a more significant association with the internal identity. Isolation turns into a hallowed space for reflection, thoughtfulness, and the development of internal harmony.

The inward excursion isn't without its difficulties. It requires an eagerness to defy uneasiness, to sit with the disquiet that emerges while confronting the unexplored world. It includes shedding layers of molding, breaking liberated from the shackles of cultural assumptions, and embracing the weakness that accompanies validness.

In the midst of the difficulties, the inward excursion is likewise a wellspring of flexibility. It gives a safe-haven of solidarity that rises above outside conditions. The bits of knowledge acquired through self-reflection become a wellspring of internal power, empowering people to explore the tempests of existence with effortlessness and backbone.

Connections, both with oneself and with others, assume a crucial part in the internal excursion. The manner in which people connect with themselves establishes the vibe for how they draw in with the world. Self-sympathy, confidence, and self-acknowledgment become foundations of a solid relationship with oneself.

In the domain of relational associations, the inward excursion encourages a more profound comprehension of others. Compassion, conceived out of mindfulness, turns into a scaffold that interfaces people on a significant level. The capacity to see past superficial communications and fathom the common human experience reinforces the texture of connections.

The internal excursion is likewise a journey of feelings. It includes embracing the full range of sentiments — from happiness to distress, from affection to fear. Every inclination is a courier, conveying significant bits of knowledge about the inward scene. Figuring out how to sit with uneasiness and permitting feelings to stream without obstruction is a critical part of the capacity to understand individuals on a deeper level developed on the internal excursion.

Otherworldliness frequently entwines with the internal excursion, in spite of the fact that it takes assorted structures for various people. As far as some might be concerned, otherworldliness is inseparable from coordinated religion, giving a structure to interfacing with the heavenly. For other people, otherworldliness is a more private investigation, a journey for greatness that goes past strict limits.

The inward excursion unfurls right now, yet it is additionally unpredictably associated with the embroidery of time. Considering the past turns into an instrument for figuring out designs, perceiving repeating subjects, and removing insight from encounters. All the while, the internal excursion welcomes people to mull over the future, imagining the individual they seek to turn into.

The idea of time takes on an alternate aspect in the internal excursion. It isn't simply a straight movement however a repeating dance of development, change, and recharging. The past, present, and future merge in the timeless now, where every second turns into a material for self-articulation and self-development.

The inward excursion is a course of unsuitable and becoming — a shedding of old skins and a resurrection into legitimacy. It is an excursion of coordination, where divided parts of oneself are embraced and orchestrated. The internal speculative chemistry that happens permits people to take advantage of their maximum capacity and carry on with a daily existence lined up with their most profound qualities.

At last, the inward excursion is a ceaseless mission. It is a dynamic, developing cycle that unfurls in layers, uncovering new profundities and bits of knowledge with each turn. An excursion of self-revelation stretches out past the limits of the individual, adding to the aggregate embroidery of human cognizance.

All in all, the inward excursion is a consecrated odyssey into the profundities of oneself. An extraordinary journey requires mental fortitude, thoughtfulness, and a readiness to face the intricacies of the human experience. Through self-revelation, care, and a profound association with others, people set out on an excursion of credibility, reason, and significant importance. The inward excursion is an investigation of the spirit — an everlasting journey that improves the embroidery of human life.

4.1 Delving into the concept of the inner journey in Theorophy.

The idea of the internal excursion holds a focal spot in Theorophy, a philosophical structure that envelops both hypothetical comprehension and otherworldly insight. Theorophy, frequently alluded to as the insight of God, investigates the interconnectedness of the heavenly, the universe, and the human experience. Inside this far reaching system, the internal excursion arises as a vital road for people to develop how they might interpret self, the universe, and the heavenly.

Theorophy places that the inward excursion isn't simply a mental or individual investigation however a fundamental piece of an inestimable embroidery. It proposes that by diving into the profundities of one's being, people can uncover the all inclusive bits of insight that associate them to the more prominent entirety. Theorophy welcomes people to set out on a groundbreaking mission, an excursion that rises above the limits of conventional discernment and wanders into the enchanted domains of presence.

At the core of Theorophy lies the acknowledgment that the material and profound aspects are entwined, each impacting and forming the other. The inward excursion, consequently, turns into a method for orchestrating these aspects, adjusting individual development to the infinite request. A course of self-acknowledgment goes past the superficial parts of personality, diving into the immortal and vast nature of the spirit.

In the Theorophical understanding, the inward excursion starts with mindfulness — a cognizant acknowledgment of the different layers of oneself and an affirmation of the heavenly flash inside. This mindfulness isn't established in egoic ID however in a more profound, more significant comprehension of one's otherworldly embodiment. Theorophy instructs that by turning internal, people can get to the heavenly insight that dwells inside the center of their being.

The internal excursion in Theorophy includes a persistent course of purging and refinement. It is similar to the catalytic change of base metals into gold, representing the height of the human soul toward its heavenly potential. The refinement cycle isn't tied in with dismissing or stifling parts of oneself yet about rising above constraints and lining up with the heavenly diagram encoded inside the spirit.

Vital to Theorophical lessons is the possibility that the inward excursion is a way of climb — a rising towards higher conditions of cognizance and profound mindfulness. An excursion rises above the restrictions of the material world, permitting people to interface with the unpretentious domains of presence. Theorophy accentuates the development of ideals, moral lead, and otherworldly practices as fundamental components in this climb.

Reflection, thought, and petitioning heaven are necessary parts of the inward excursion in Theorophy. These practices act as entryways to higher conditions of awareness, empowering people to collective with the heavenly and get direction from the otherworldly domains. Theorophy instructs that through standard profound practices, people can adjust themselves to the enormous harmonies and partake in the heavenly dance of creation.

Theorophical cosmology sets that the internal excursion is definitely not a single undertaking yet a participatory demonstration in the grandiose ensemble. It imagines a universe imbued with divine knowledge, where each being is interconnected and assumes an extraordinary part in the unfurling of the heavenly arrangement. The inward excursion, hence, turns into a cognizant joint effort with the enormous powers, an agreeable dance of co-creation.

Theorophy additionally investigates the idea of heavenly characteristic — the possibility that the heavenly isn't far off or isolate from the world however penetrates each part of creation. This understanding changes the internal excursion into a consecrated investigation of the heavenly inside the commonplace. It welcomes people to perceive the presence of the heavenly in each experience, relationship, and articulation of life.

As people progress on the inward excursion in Theorophy, they might experience difficulties and tests that act as any open doors for development and change. These difficulties are seen not as impediments but rather as impetuses for otherworldly advancement. Theorophy instructs that misfortune is an educator, directing people to more profound bits of knowledge, versatility, and a more prominent comprehension of the heavenly reason.

Theorophical insight proposes that the inward excursion is a recurrent course of death and resurrection — a consistent shedding of the old and a rise into the new. This repetitive nature reflects the vast patterns of creation, protection, and disintegration. Theorophy urges people to embrace the fleetingness of the material world and to secure their cognizance in the everlasting and perpetual substance.

The internal excursion in Theorophy is likewise a way of self-greatness. It includes rising above the restricted feeling of individual personality and growing attention to incorporate the general Self. Theorophical lessons draw from mysterious customs that talk about the unity of all presence, stressing the interconnectedness of each being with the heavenly source.

In Theorophical reasoning, the internal excursion is indivisible from the idea of adoration. Love is viewed as the bringing together power that ties all of creation together. The inward excursion turns into an excursion of affection — an extending and extension of the heart's ability to genuinely cherish. Theorophy instructs that genuine affection exudes from the heavenly source and courses through people, associating them with the whole universe.

Theorophy urges people to develop temperances like sympathy, thoughtfulness, and absolution as they progress on the internal excursion. These ideals are seen as moral standards as well as articulations of heavenly characteristics intrinsic in the spirit. Theorophy instructs that by typifying these ideals, people fall in line with the heavenly traits and add to the upliftment of the shared perspective.

The inward excursion in Theorophy is additionally personally associated with the idea of instinct. Instinct is seen as the immediate realizing that emerges from the spirit's association with the heavenly. Theorophical lessons stress the significance of tuning into instinctive direction as people explore the intricacies of life. Instinct turns into a compass that adjusts people to the heavenly will and leads them on the way of honesty.

Theorophical shrewdness proposes that the internal excursion finishes in a condition of profound arousing — an enlivening to the real essence of the real world and the heavenly reason behind presence. This enlivening is definitely not a one-time situation however a nonstop transpiring of cognizance. Theorophy instructs that as people stir to higher conditions of mindfulness, they add to the aggregate arousing of mankind and the whole universe.

Theorophy imagines an existence where people, directed by the insight of the internal excursion, work together in the co-formation of an amicable and illuminated society. Theorophical lessons underscore the significance of administration to others as a characteristic articulation of profound acknowledgment. The internal excursion, when lived legitimately, moves people to contribute benevolently to the prosperity of the world.

All in all, Theorophy presents a significant and all encompassing comprehension of the internal excursion. It goes past the limits of individual brain research and coordinates otherworldly insight with infinite experiences. Theorophy imagines the inward excursion as a groundbreaking mission — a sacrosanct journey that leads people from mindfulness to self-acknowledgment, from individual development to grandiose arrangement. The internal excursion, with regards to Theorophy, is a sacrosanct hit the dance floor with the heavenly — an excursion of climb, love, and administration that adds to the vast orchestra of creation.

4.2 Discussing meditation, contemplation, and self-discovery as essential practices.

Reflection, consideration, and self-disclosure stand as points of support in the domain of individual and otherworldly turn of events, offering significant roads for people to investigate the profundities of their cognizance and open the secrets of their internal identities. These practices, frequently interweaved and integral, give a system to self-investigation, profound recuperating, and otherworldly development. As we dig into the meaning of reflection, thought, and self-disclosure, it becomes clear that they are not simple instruments but rather groundbreaking cycles that work with an agreeable association between the psyche, body, and soul.

Contemplation, in its bunch structures, fills in as a door to inward quietness and elevated mindfulness. At its center, contemplation is the deliberate demonstration of centering the psyche, frequently through the breath or a particular place of consideration, to accomplish a condition of mental lucidity and serenity. Whether established in old profound customs or current mainstream rehearses, contemplation shares a consistent idea — developing care. Care, in this specific situation, alludes to the non-critical consciousness of one's viewpoints, feelings, and sensations right now.

Care contemplation, as advocated in contemporary settings, includes pointing out the breath, noticing considerations without connection, and establishing oneself in the present. This training urges people to become observers to their internal encounters, cultivating a feeling of separation from the steady stream of considerations that frequently rules the brain. Through steady reflection, people foster the ability to notice the changes of the brain and, in doing as such, gain a more profound comprehension of their thinking designs and close to home reactions.

The advantages of contemplation reach out past the psychological domain, affecting actual prosperity. Research has demonstrated the way that ordinary reflection practice can prompt decreases in pressure, tension, and side effects of wretchedness. Moreover, it has been related with upgrades in focus, mental capability, and in general close to home strength. The physiological impacts of reflection, for example, brought down pulse and diminished cortisol levels, highlight its all encompassing effect on the brain body association.

Examination, while imparting similitudes to contemplation, includes a more engaged and intelligent commitment with explicit considerations, ideas, or questions. Dissimilar to the calm quietness of reflection, thought is a functioning cycle that welcomes people to contemplate, ask, and investigate the more profound components of their reality. Thoughtful practices might include the investigation of philosophical or otherworldly lessons, the assessment of individual qualities, or the consideration of life's crucial inquiries.

One striking type of pensive practice is intelligent journaling, where people record their contemplations, sentiments, and experiences. This course of externalizing inward encounters through putting down gives a substantial account of the insightful excursion, permitting people to return to and consider their developing comprehension of self and the world. Consideration, fundamentally, is a cognizant commitment with the intricacies of human experience, welcoming people to explore the domains of importance, reason, and existential request.

Thoughtful practices can likewise take on a more creative structure, like through visual workmanship, music, or dance. The inventive strategy turns into a mechanism for self-articulation and investigation, permitting people to take advantage of the instinctive and emblematic parts of their cognizance. Through imaginative examination, people might reveal stowed away parts of themselves and gain bits of knowledge that rise above verbal enunciation.

The connection among reflection and consideration becomes obvious in the collaboration of these practices. Reflection gives the quietness and lucidity fundamental for thought, while consideration advances the profundity of contemplation by carrying cognizant investigation into the reflective space. Together, they make a unique collaboration that encourages mindfulness, scholarly investigation, and a significant association with the internal identity.

Self-revelation, as a more extensive idea, envelops both reflection and examination while reaching out into the texture of day to day existence. It is a continuous course of uncovering and grasping the layers of one's character, convictions, and inspirations. Self-disclosure includes an eagerness to investigate the shadow perspectives — the covered up and once in a while testing features of the self that might incorporate unsettled feelings, fears, or restricting convictions.

Reflection and consideration act as integral assets in the excursion of self-disclosure.

Through reflection, people foster the capacity to notice their contemplations and feelings without judgment, making a space for the divulging of oblivious examples. Thought, then again, gives an organized system to effectively looking at the layers of individual personality and diving into the basic suspicions that shape one's viewpoint.

One part of self-revelation is the acknowledgment of the inner self — the built identity that is molded by cultural impacts, previous encounters, and social molding. Reflection and thought assume a urgent part in destroying the egoic structures, permitting people to see the pith of their being past the layers of recognizable proof. As the egoic connections release, a more bona fide and freed identity arises.

The excursion of self-revelation includes embracing weakness and validness. It requires a readiness to confront awkward bits of insight, challenge deliberate constraints, and explore the vulnerabilities of self-improvement. Reflection and thought, as fundamental parts of this excursion, give the devices to self-request and the development of mindfulness.

Care, a vital component in both reflection and thought, is a focal part of self-disclosure. By developing care, people foster the ability to notice their contemplations, feelings, and ways of behaving with an increased feeling of mindfulness. This mindfulness turns into a lamp that enlightens the way of self-revelation, exposing the complexities of the inward scene.

Self-revelation is a continuous, transformative cycle that unfurls all through the different phases of life. It includes a constant reconsideration of one's convictions, values, and needs. The incorporation of contemplation and thought into day to day existence upholds this cycle by cultivating an intelligent disposition and a responsive receptiveness to the steadily unfurling nature of oneself.

With regards to reflection, care practices, for example, body check contemplations, adoring graciousness contemplations, or careful development add to the development of mindfulness. These practices urge people to bring a non-critical and empathetic attention to their actual sensations, feelings, and relational connections. From the perspective of care, self-revelation turns into an excursion of presence and acknowledgment.

Thoughtful practices in self-disclosure might include investigating individual accounts, analyzing center convictions, and examining presumptions concerning personality and reason. Journaling, as a scrutinizing instrument, gives an organized space to people to dig into the subtleties of their inward world, revealing bits of knowledge that might have been darkened by the commotion of day to day existence. Consideration turns into a mirror mirroring the inward scenes of oneself.

The convergence of reflection, examination, and self-disclosure turns out to be especially strong when people participate in purposeful and coordinated rehearses. For instance, a pensive reflection meeting might include considering explicit subjects or inquiries prior to entering a quiet contemplation, permitting the bits of knowledge acquired through examination to develop during the thoughtful state. This coordinated methodology boosts the synergistic advantages of the two practices.

As people progress on the way of self-disclosure, they might experience snapshots of significant knowledge, epiphanies that reshape how they might interpret themselves and the world. These bits of knowledge might emerge immediately during reflection, consideration, or even amidst regular exercises. They act as signs on the excursion, directing people toward more noteworthy mindfulness and a more genuine approach to being.

The course of self-revelation isn't without its difficulties. It expects fortitude to face parts of the self that might be awkward or new. Reflection and thought, as partners in this excursion, give the essential apparatuses to explore these difficulties with poise and strength. The act of self-empathy turns into a vital buddy, permitting people to embrace their flaws and commend their novel process of self-revelation.

Taking everything into account, reflection, thought, and self-disclosure structure an indivisible ternion that winds around together the texture of individual and otherworldly development. Reflection makes the way for inward tranquility and care, examination offers an organized investigation into the profundities of oneself, and self-revelation is the continuous excursion of revealing the layers of personality. Together, these practices make an extraordinary collaboration that engages people to explore the intricacies of the internal scene with effortlessness, legitimacy, and significant mindfulness.

4.3 Exploring the transformative nature of the inner journey and its impact on personal growth.

The internal excursion, a groundbreaking odyssey into the profundities of oneself, is a significant and dynamic cycle that shapes the course of self-awareness. A journey rises above the outer features of life and dives into the center of one's being, unwinding layers of character, convictions, and encounters. The extraordinary idea of the inward excursion lies in its ability to catalyze self-revelation, cultivate strength, and light a ceaseless pattern of development and development.

At the core of the groundbreaking inward excursion is the acknowledgment that self-improvement is certainly not a straight movement however a multi-faceted investigation. It includes stripping back the layers of molded reactions, cultural assumptions, and acquired convictions to uncover the true self underneath. The extraordinary perspective arises as people defy their feelings of dread, embrace weakness, and take part in a cognizant course of becoming.

The inward excursion unfurls against the scenery of regular day to day existence, yet a journey requires purposeful and contemplative commitment. It is a takeoff from the outer quests for progress and acknowledgment to the inside scenes of mindfulness and self-acknowledgment. The extraordinary idea of this excursion becomes obvious as people navigate the territory of their own mind, experiencing both the shadows and the light that shape their self-improvement.

A focal topic in the groundbreaking inward excursion is the course of self-disclosure. This includes looking into the profundities of one's awareness, addressing suspicions, and investigating the complexities of individual personality. Self-revelation is definitely not a one-time occasion yet a continuous investigation, a pledge to uncovering the layers that have collected over the long run. The extraordinary effect lies in the freshly discovered understanding that emerges from this investigation — a more profound consciousness of one's qualities, wants, and true self.

The extraordinary internal excursion frequently starts with an impetus — a snapshot of enlivening, a groundbreaking occasion, or an acknowledgment that prompts people to leave on a way of self-investigation. This impetus fills in as a flash that touches off

the longing for self-awareness and gets the groundbreaking excursion rolling. It very well may be a critical life progress, for example, a profession change, a relationship shift, or an individual emergency that prompts people to rethink their lives and look for more profound significance.

As people adventure into the extraordinary internal excursion, they experience the landscape of mindfulness. This is the most common way of turning internal, noticing considerations and feelings without judgment, and developing a careful presence.

The groundbreaking effect of mindfulness is significant — it makes the way for cognizant decision and enables people to break liberated from programmed responses and adapted reactions. Through mindfulness, people gain knowledge into their thinking designs, profound triggers, and ongoing ways of behaving, establishing the groundwork for deliberate self-improvement.

The groundbreaking idea of the inward excursion is likewise woven into the texture of the ability to appreciate anyone on a deeper level. The capacity to appreciate anyone on a deeper level includes perceiving, understanding, and dealing with one's own feelings, as well as exploring the feelings of others. As people participate in the inward excursion, they foster an uplifted profound mindfulness, figuring out how to embrace the full range of their sentiments. This groundbreaking system takes into consideration a more bona fide articulation of feelings and cultivates profound versatility — a fundamental part of self-improvement.

Weakness is a vital component in the extraordinary internal excursion. It includes the ability to open oneself to vulnerability, risk, and close to home openness. Embracing weakness is groundbreaking since it destroys the defensive walls people might have worked around themselves. By permitting themselves to be seen legitimately, people make space for association, sympathy, and authentic connections, encouraging self-awareness with regards to significant associations with others.

The groundbreaking effect of the inward excursion is apparent during the time spent delivering restricting convictions. These are imbued contemplations and suppositions that might have been procured over the course of life and have become boundaries to self-improvement. The internal excursion welcomes people to challenge and destroy these restricting convictions, opening up additional opportunities and growing the limits of what they accept is feasible. This groundbreaking change in outlook makes ready for expanded fearlessness and a more extensive point of view on private potential.

Absolution is one more groundbreaking aspect of the internal excursion. The eagerness to pardon oneself as well as other people delivers the grasp of disdain and permits people to push ahead unburdened by the heaviness of past complaints. Pardoning isn't tied in with supporting hurtful activities; rather, a strong demonstration of self-freedom adds to self-awareness by opening up energy and mental space for positive change.

The extraordinary internal excursion is set apart by an extending healthy identity empathy. Self-sympathy includes treating oneself with graciousness and grasping, particularly despite difficulties and mishaps. The extraordinary effect of self-sympathy lies in its capacity to neutralize self-analysis and encourage a supporting inside climate. This empathetic position toward oneself turns into a strong starting point for self-awareness, empowering people to gain from missteps and misfortunes as opposed to being incapacitated by them.

As people progress on the extraordinary internal excursion, they frequently experience the shadows — the parts of themselves that they might have stifled or denied. This showdown with the shadows is a fundamental stage in self-awareness. It includes recognizing and coordinating the more obscure parts of oneself, perceiving that completeness incorporates both light and shadow. The extraordinary effect lies in the freedom that comes from tolerating all features of oneself, prompting a more coordinated and true identity.

The extraordinary idea of the internal excursion is personally associated with the idea of care. Care, established in old thoughtful practices, includes developing a non-critical consciousness of the current second. The groundbreaking effect of care lies in its capacity to moor people in the present time and place, diminishing pressure, upgrading lucidity, and encouraging a more profound association with the unfurling excursion of self-improvement. Through care, people foster the ability to answer life's difficulties with more prominent composure and strength.

The groundbreaking internal excursion likewise incorporates the investigation of values and reason. People take part in an intelligent cycle to observe their fundamental beliefs — those core values that characterize what is significant and important to them. This investigation of values fills in as a compass for self-awareness, directing people toward decisions and activities that line up with their real selves. The groundbreaking effect lies in the arrangement of one's existence with profoundly held values, cultivating a feeling of direction and satisfaction.

Otherworldliness frequently interweaves with the groundbreaking internal excursion. Whether established in strict customs or a more common feeling of association with the universe, otherworldliness gives a structure to people to investigate existential inquiries and develop a more profound feeling of significance. The groundbreaking effect of otherworldliness lies in its capacity to rise above the egoic self and associate people with an option that could be more significant than themselves. This feeling of greatness encourages a viewpoint that goes past the transient difficulties of life, adding to strength and a more extensive setting for self-awareness.

Connections assume an essential part in the groundbreaking inward excursion. The manner in which people connect with themselves as well as other people turns into a mirror mirroring their development and self-revelation. The groundbreaking effect of cognizant connections lies in the common help and shared development that arises when people connect legitimately with one another. Solid connections become a supporting ground for self-improvement, giving open doors to reflection, criticism, and the developing of the capacity to understand individuals on a profound level.

The extraordinary internal excursion is a continuous cycle instead of an objective. It includes patterns of self-reflection, learning, and transformation. The extraordinary effect isn't just private yet in addition stretches out to the more extensive setting of cultural and aggregate development. As people go through private change, they add to an expanding influence, impacting the networks and frameworks they are a piece of.

All in all, the extraordinary idea of the inward excursion is an embroidery woven with strings of mindfulness, the capacity to understand people on a profound level, weakness, pardoning, self-sympathy, and care. As people explore the territory of self-disclosure, stand up to restricting convictions, and incorporate the shadows, they go through a significant change that waves through their connections, their networks, and the bigger embroidery of human cognizance. The inward excursion, with its extraordinary effect on self-improvement, is an encouragement to embrace the unique course of turning into one's generally valid self — an excursion that unfurls in every second and reverberates with the immortal reverberations of human potential.

The inward excursion, a journey into the significant profundities of oneself, is an extraordinary odyssey that applies a significant effect on self-awareness. It unfurls as a complicated embroidery, winding around together the strings of self-disclosure, strength, and consistent development. Dissimilar to outer pursuits, the inward excursion is a visit into the domains of cognizance, uncovering layers of character, convictions, and encounters. Its extraordinary nature lies in its capacity to catalyze mindfulness, sustain the capacity to understand people on a profound level, and light an unending pattern of development and development.

At its center, the internal excursion is a course of self-revelation. It includes stripping away the layers that have gathered over the long haul, uncovering the credible self underneath the facade of cultural assumptions and adapted reactions. This cycle is certainly not a direct movement yet a multi-layered investigation, a promise to revealing the layers that shape one's character. The groundbreaking effect of self-revelation becomes clear in the newly discovered understanding that emerges — a more profound consciousness of one's qualities, wants, and legitimate self.

The commencement of the internal excursion frequently starts with an impetus — a snapshot of enlivening, a huge life altering situation, or an acknowledgment that moves people to leave on a way of self-investigation. This impetus fills in as a flash that touches off the craving for self-improvement and gets the groundbreaking excursion under way. It very well may be a profession shift, a relationship progress, or an individual emergency that prompts people to rethink their lives and look for more profound significance.

As people set out on the groundbreaking inward excursion, they explore the landscape of mindfulness. This includes turning internal, noticing contemplations and feelings without judgment, and developing a careful presence. The extraordinary effect of mindfulness lies in its capacity to make the way for cognizant decision. It engages people to break liberated from programmed responses and molded reactions, giving the establishment to purposeful self-awareness.

The extraordinary idea of the inward excursion is additionally woven into the texture of the ability to appreciate anyone on a profound level. The capacity to appreciate people on a deeper level incorporates the acknowledgment, understanding, and the board of one's own feelings, as well as the capacity to explore the feelings of others.

As people participate in the inward excursion, they foster elevated close to home mindfulness, figuring out how to embrace the full range of their sentiments. This extraordinary interaction considers a more true articulation of feelings and encourages close to home versatility — a fundamental part of self-improvement.

Weakness arises as a vital component in the extraordinary internal excursion. It requires the eagerness to open oneself to vulnerability, risk, and profound openness. Embracing weakness is extraordinary in light of the fact that it destroys the defensive walls people might have worked around themselves. By permitting themselves to be seen legitimately, people make space for association, compassion, and certifiable connections, cultivating self-awareness inside the setting of significant associations with others.

Restricting convictions address hindrances to self-awareness, and the inward excursion fills in as a vehicle for their delivery. These convictions are instilled contemplations and suspicions that might have been obtained over the course of life and have become deterrents to self-awareness. The internal excursion welcomes people to challenge and destroy these restricting convictions, opening up additional opportunities and growing the limits of what they accept is reachable. This extraordinary change in mentality prepares for expanded fearlessness and a more extensive point of view on private potential.

The extraordinary effect of the inward excursion is interwoven with the idea of pardoning. The ability to excuse oneself as well as other people delivers the hold of disdain and permits people to push ahead unburdened by the heaviness of past complaints. Pardoning isn't tied in with overlooking hurtful activities; rather, a strong demonstration of self-freedom adds to self-awareness by opening up energy and mental space for positive change.

Self-sympathy is one more extraordinary feature of the internal excursion. It includes treating oneself with generosity and grasping, particularly despite difficulties and misfortunes. The groundbreaking effect of self-empathy lies in its capacity to check self-analysis and cultivate a supporting inside climate. This sympathetic position toward oneself turns into a strong starting point for self-awareness, empowering people to gain from slip-ups and misfortunes as opposed to being deadened by them.

As people progress on the groundbreaking internal excursion, they frequently experience the shadows — the parts of themselves that they might have smothered or denied. This showdown with the shadows is a fundamental stage in self-awareness. It includes recognizing and coordinating the hazier parts of oneself, perceiving that completeness incorporates both light and shadow. The groundbreaking effect lies in the freedom that comes from tolerating all features of oneself, prompting a more coordinated and genuine identity.

The groundbreaking idea of the internal excursion is personally associated with the idea of care. Care, established in old pondering customs, includes developing a non-critical consciousness of the current second. The groundbreaking effect of care lies in its capacity to secure people in the present time and place, diminishing pressure, upgrading clearness, and cultivating a more profound association with the unfurling excursion of self-improvement. Through care, people foster the ability to answer life's difficulties with more noteworthy serenity and flexibility.

The inward excursion likewise includes the investigation of values and reason. People take part in an intelligent cycle to observe their basic beliefs — core values that characterize what is significant and important to them. This investigation of values fills in as a compass for self-awareness, directing people toward decisions and activities that line up with their genuine selves. The extraordinary effect lies in the arrangement of one's existence with profoundly held values, cultivating a feeling of direction and satisfaction.

Otherworldliness frequently interlaces with the groundbreaking inward excursion, giving a structure to people to investigate existential inquiries and develop a more profound feeling of importance. Whether established in strict customs or a more common feeling of association with the universe, otherworldliness rises above the egoic self and interfaces people with an option that could be more significant than themselves. This feeling of greatness cultivates a viewpoint that goes past the transient difficulties of life, adding to flexibility and a more extensive setting for self-awareness.

Connections assume a crucial part in the extraordinary inward excursion. The manner in which people connect with themselves as well as other people turns into a mirror mirroring their development and self-disclosure. Cognizant connections add to self-awareness, giving open doors to reflection, criticism, and the developing of the capacity to appreciate individuals on a profound level. Solid connections become a sustaining ground for self-improvement, encouraging common help and shared development.

The groundbreaking internal excursion is a continuous cycle as opposed to an objective. It includes patterns of self-reflection, learning, and transformation. The extraordinary effect isn't just private yet reaches out to the more extensive setting of cultural and aggregate development. As people go through private change, they add to an expanding influence, impacting the networks and frameworks they are a piece of.

All in all, the groundbreaking idea of the inward excursion is an embroidery woven with strings of mindfulness, the capacity to understand people on a profound level, weakness, pardoning, self-sympathy, and care. As people explore the landscape of self-disclosure, stand up to restricting convictions, and incorporate the shadows, they go through a significant change that waves through their connections, their networks, and the bigger embroidery of human cognizance.

The internal excursion, with its extraordinary effect on self-improvement, is an encouragement to embrace the powerful course of turning into one's generally valid self — an excursion that unfurls in every second and resounds with the immortal reverberations of human potential.

Chapter 5

Theorophical Ethics

Hypothetical morals, otherwise called regularizing morals, is a part of reasoning that looks to comprehend and assess the ethical rules that guide human way of behaving. This field dives into the idea of profound quality, looking at inquiries regarding what is correct or off-base, positive or negative, and how people should act in different circumstances. Hypothetical morals plans to give a structure to moral direction and to investigate the major rules that underlie moral decisions.

One of the unmistakable speculations inside hypothetical morals is deontology. Deontological morals, got from the Greek word "deon" meaning obligation, affirms that there are objective moral obligations that people are committed to follow, regardless of the outcomes. Immanuel Kant, a critical figure in deontological morals, contended that ethical activities are those directed by general rules that can be applied reliably across various circumstances. As per Kant, people have an ethical obligation to act such that regards the inborn poise of every single individual and with comply to rules that could be willed as an all inclusive regulation.

As opposed to deontology, consequentialism is another major moral hypothesis that spotlights on the results or outcomes of activities. Utilitarianism, a type of consequentialism, sets that the profound quality of a not entirely settled by the general joy or joy it produces. Jeremy Bentham and John Stuart Factory, compelling defenders of utilitarianism, contended that people ought to look for the best bliss for the best number. This approach empowers a math of delight and agony to assess the outcomes of activities and pursue moral choices in light of expanding generally speaking prosperity.

Ethicalness morals offers an alternate point of view by stressing the improvement of idealistic person characteristics. Established in antiquated Greek way of thinking, especially crafted by Aristotle, uprightness morals proposes that ethical way of behaving emerges from developing excellencies like boldness, trustworthiness, and empathy. Instead of zeroing in on rules or results, goodness morals focuses on the sort of individual one ought to turn into. Highminded people, as per this hypothesis, pursue ethically ideal choices intuitively in light of the fact that their personality has been formed by the ongoing act of temperances.

These three significant moral speculations — deontology, consequentialism, and prudence morals — give various structures to understanding and surveying moral standards. Each approach has its assets and shortcomings, and thinkers keep on discussing the benefits of these hypotheses in the journey for a far reaching and sound moral framework.

Deontological morals, with its accentuation on the job and all inclusive standards, has been evaluated for its expected unbending nature and failure to represent the intricacy of genuine circumstances. Pundits contend that severe adherence to moral standards might prompt ethically frightful results in specific conditions. For instance, a deontologist could contend that lying is in every case ethically off-base, even in circumstances where coming clean could prompt mischief. This absence of adaptability has provoked a few thinkers to investigate changes of deontological rules that integrate additional background information explicit contemplations.

Consequentialism, especially utilitarianism, faces difficulties connected with the measurement of joy and the trouble of foreseeing every one of the outcomes of an activity. The utilitarian analytics of boosting joy and limiting torment is censured for its improvement of complicated moral issues. Furthermore, pundits contend that utilitarianism might legitimize ethically problematic activities assuming they bring about a net expansion in general satisfaction. This has driven a few savants to propose varieties of consequentialism that think about factors past simple joy, like the significance of individual freedoms or the characteristic worth of specific products.

Righteousness morals, while adulated for its emphasis on character improvement, is frequently scrutinized for its absence of clear direction in moral navigation. The hypothesis gives a structure to developing excellencies however doesn't offer explicit standards or standards for figuring out what is ethically correct or wrong specifically circumstances. Pundits contend that excellence morals might be excessively emotional and not entirely clear, making it trying to determine moral conflicts. Some contemporary uprightness ethicists answer these worries by investigating ways of coordinating ethicalness morals with components of deontology or consequentialism.

Notwithstanding these major moral speculations, there are different points of view and approaches inside hypothetical morals. Contractualism, for example, investigates profound quality from the perspective of common agreements or arrangements among reasonable people. Contractualist hypotheses, created by logicians like T. M. Scanlon, propose that ethical standards are gotten from the speculative assent of reasonable specialists took part in a fair and unprejudiced dealing process.

Women's activist morals is another significant and advancing region inside hypothetical morals that inspects the effect of orientation on moral hypothesis and practice. Women's activist ethicists scrutinize conventional moral hypotheses for their verifiable disregard of ladies' points of view and encounters. They underscore the significance of resolving issues connected with orientation correspondence, care morals, and the convergences of orientation with other social classes.

The morals of care, related with women's activist morals, centers around the ethical meaning of connections and the obligations that emerge from really focusing on others. Created by scholars like Tune Gilligan, care morals challenges the customary accentuation on unique standards and features the significance of sympathy, empathy, and mindfulness of the necessities of others. This approach has been compelling in reshaping conversations about moral schooling and the job of feelings in moral direction.

Ecological morals is one more arising field inside hypothetical morals that tends to the ethical components of human communications with the climate. As worries about environmental change, biodiversity misfortune, and asset consumption heighten, ecological ethicists investigate the moral obligations people and social orders have toward the regular world. This point of view difficulties anthropocentrism — the view that human interests outweigh those of different species — and advocates for a more comprehensive and naturally feasible ethic.

Strict and social viewpoints likewise assume a huge part in molding hypothetical morals. Numerous moral hypotheses have establishes in strict practices, and strict morals keep on affecting moral talk. For instance, the Ten Precepts in Judaism and Christianity give a bunch of moral rules in light of heavenly power. Islamic morals comparatively draws on the lessons of the Quran and the Hadith to direct moral lead. Eastern philosophical customs, for example, Confucianism and Buddhism, offer one of a kind viewpoints on profound quality that vary from Western methodologies.

As hypothetical morals investigates these different points of view, it wrestles with essential inquiries concerning the idea of profound quality itself. Metaethics, a subfield of hypothetical morals, looks at the otherworldly and epistemological parts of ethical quality. Logicians in this field research the idea of moral truth, the significance of moral language, and the reason for moral cases. Metaethical questions incorporate whether moral realities exist freely of human convictions, whether moral explanations express genuine insights, and how people come to be aware or find moral standards.

The connection among morals and human brain research is one more area of request inside hypothetical morals. Moral brain science investigates how people foster moral convictions, make moral decisions, and take part in moral way of behaving. This interdisciplinary field draws on bits of knowledge from brain research, neuroscience, and reasoning to comprehend the mental cycles and profound elements that shape moral direction. For instance, research in moral brain science plays analyzed the part of sympathy, moral instincts, and the impact of social and social variables on moral turn of events.

Moral relativism is a metaethical position that declares the variety of moral convictions across societies and people. As indicated by moral relativism, there is no goal or general norm for deciding the accuracy of moral decisions. All things considered, moral standards are viewed as comparative with the social or individual setting in which they emerge.

While moral relativism recognizes the significance of social variety and verifiable setting, it likewise raises difficulties connected with the potential for moral advancement and the capacity to scrutinize unsafe practices inside a specific culture.

The continuous discussions and advancements inside hypothetical morals feature the dynamic and developing nature of moral request. As new difficulties arise in regions like biotechnology, man-made brainpower, and worldwide equity, ethicists wrestle with the utilization of existing hypotheses to novel moral issues. For instance, issues encompassing the utilization of hereditary designing to adjust human characteristics bring up issues about the moral limits of mechanical mediations in human instinct and the expected ramifications for social equity.

The moral ramifications of arising innovations, like independent vehicles and high level observation frameworks, additionally brief moral reflection. Savants and ethicists take part in conversations about the capable turn of events and arrangement of these advancements, taking into account issues connected with security, independence, and the potential for unseen side-effects. The convergence of morals and innovation highlights the requirement for progressing moral request to address the moral difficulties of our quickly influencing world.

The job of useful thinking in moral direction is a focal topic in hypothetical morals. Viable thinking includes the course of consideration and judgment through which people explore moral predicaments and settle on decisions that line up with their moral standards. Speculations of pragmatic thinking investigate the mental cycles engaged with weighing contending moral contemplations, surveying the importance of moral standards to explicit cases, and showing up at ethically legitimized choices.

Moral consultation frequently includes adjusting contending values and taking into account the points of view of various partners. The idea of moral pluralism recognizes the presence of different, possibly clashing, moral rules that people should explore. Moral pluralism features the intricacy of moral navigation and underscores the significance of cautious reflection on the subtleties of specific circumstances. In tending to moral contentions, people might have to focus on specific qualities over others or look for a fair methodology that regards different moral contemplations.

The advancement of moral person is one more key part of hypothetical morals. Prudence ethicists, specifically, stress the significance of developing righteous qualities that add to moral greatness. This course of character improvement includes the routine act of ideals like genuineness, uprightness, fortitude, and empathy. Moral schooling and moral development assume a critical part in molding people's characters and impacting their ability for moral way of behaving.

The topic of moral obligation is a repetitive subject in hypothetical morals. Hypotheses of moral obligation investigate the circumstances under which people are considered ethically responsible for their activities. Deciding moral obligation includes considering variables like goal, information, and the capacity to go with independent decisions. Rationalists additionally analyze the ramifications of moral obligation regarding issues like law enforcement, discipline, and cultural assumptions for responsibility.

The connection among morals and regulation is one more area of crossing point, with hypothetical morals giving an establishment to the turn of events and assessment of lawful standards. Legitimate positivism, a way of thinking of regulation, declares that the authenticity of lawful guidelines isn't dependent upon their ethical rightness. Notwithstanding, the association among regulation and morals stays a complex and discussed subject, with progressing conversations about the ethical legitimacy of specific regulations and the job of moral contemplations in lawful direction.

Hypothetical morals likewise draws in with inquiries of civil rights, resolving issues of decency, correspondence, and the appropriation of assets inside social orders. Hypotheses of equity, for example, John Rawls' hypothesis of equity as decency, propose standards for coordinating social organizations to guarantee an only dispersion of advantages and weights. Common agreement speculations, drawing on the thoughts of logicians like Hobbes, Locke, and Rousseau, investigate the speculative arrangements that people would make to lay out an equitable and helpful society.

As well as looking at moral standards at the individual and cultural levels, hypothetical morals stretches out its request to the worldwide field. Worldwide morals investigates the ethical elements of global relations, basic freedoms, and worldwide administration. Savants in this field consider questions connected with the obligations of people, states, and worldwide associations in tending to worldwide difficulties, for example, neediness, environmental change, and furnished struggle. The idea of cosmopolitanism, which underlines a feeling of worldwide citizenship and shared honest convictions, has acquired conspicuousness in conversations about worldwide morals.

As hypothetical morals keeps on developing, contemporary logicians connect with interdisciplinary points of view and draw on bits of knowledge from fields like brain research, social science, financial matters, and political theory. This interdisciplinary methodology improves moral request by integrating observational exploration and viable contemplations into philosophical examination. The joint effort among ethicists and experts in different fields mirrors a guarantee to addressing true moral difficulties and adding to the improvement of morally informed strategies and practices.

All in all, hypothetical morals envelops a rich embroidery of viewpoints, speculations, and discussions that enlighten the intricacies of moral thinking and navigation. From deontological standards to consequentialist estimations and the development of prudent person, moral hypotheses offer assorted systems for understanding and assessing human way of behaving. As ethicists wrestle with arising difficulties and participate in interdisciplinary exchange, hypothetical morals stays a dynamic and crucial field that keeps on molding how we might interpret profound quality and guide our moral undertakings in a consistently impacting world.

5.1 Examining the ethical framework within Theorophy.

Looking at the moral system inside hypothetical way of thinking includes a far reaching investigation of the basic standards, hypotheses, and discussions that shape how we might interpret profound quality and guide moral navigation. Hypothetical morals, otherwise called regularizing morals, looks to resolve principal inquiries regarding what is correct or off-base, fortunate or unfortunate, and how people should act in different circumstances. In this assessment, we dive into key moral speculations, their studies, and the more extensive philosophical requests that add to the continuous advancement of moral systems.

Deontology, a noticeable moral hypothesis, declares that there are objective moral obligations that people are committed to follow, independent of the results. Immanuel Kant, a focal figure in deontological morals, contended that ethical activities ought to be directed by general rules that can be applied reliably across various circumstances. As per Kant, people have an ethical obligation to act such that regards the inborn nobility of every single individual and with comply to rules that could be willed as a general regulation.

Pundits of deontology contend that its accentuation on severe adherence to moral guidelines might prompt unbending nature and a failure to represent the intricacy of genuine circumstances. The hypothesis' absolutist nature, which recommends that specific activities are in every case ethically off-base paying little mind to setting, has provoked a few thinkers to investigate changes that consolidate additional background info explicit contemplations. Adjusting the obligation to all inclusive standards with an acknowledgment of the subtleties of specific circumstances stays a test inside deontological morals.

Consequentialism, another major moral hypothesis, centers around the results or outcomes of activities. Utilitarianism, a type of consequentialism, places that the profound quality of a not entirely set in stone by the general joy or delight it produces. Jeremy Bentham and John Stuart Factory, persuasive defenders of utilitarianism, contended that people ought to look for the best joy for the best number. This approach supports a math of joy and torment to assess the results of activities and go with moral choices in view of boosting generally prosperity.

Pundits of consequentialism, especially utilitarianism, raise worries about the evaluation of satisfaction and the test of foreseeing every one of the outcomes of an activity.

The utilitarian analytics of expanding joy and limiting torment is condemned for its rearrangements of intricate moral issues. Furthermore, pundits contend that utilitarianism might legitimize ethically problematic activities assuming they bring about a net expansion in by and large satisfaction. This has driven a few scholars to propose varieties of consequentialism that think about factors past simple joy, like the significance of individual freedoms or the natural worth of specific merchandise.

Righteousness morals, established in old Greek way of thinking, gives an alternate point of view by stressing the improvement of upright person characteristics. Aristotle, a vital figure in righteousness morals, contended that ethical way of behaving emerges from developing excellencies like mental fortitude, genuineness, and sympathy. Instead of zeroing in on rules or results, ideals morals focuses on the sort of individual one ought to turn into. Highminded people, as indicated by this hypothesis, pursue ethically ideal choices instinctually in light of the fact that their personality has been molded by the routine act of excellencies.

While temperance morals is lauded for its emphasis on character improvement, it is likewise reprimanded for its absence of clear direction in moral navigation. The hypothesis gives a structure to developing ideals yet doesn't offer explicit standards or standards for figuring out what is ethically correct or wrong specifically circumstances. Pundits contend that ethicalness morals might be excessively emotional and not entirely clear, making it trying to determine moral conflicts. Some contemporary righteousness ethicists answer these worries by investigating ways of incorporating uprightness morals with components of deontology or consequentialism.

Past these major moral speculations, there are different viewpoints and approaches inside hypothetical morals. Contractualism, for example, investigates ethical quality from the perspective of common agreements or arrangements among reasonable people. Contractualist speculations recommend that ethical standards are gotten from the theoretical assent of reasonable specialists took part in a fair and unbiased dealing process. T. M. Scanlon, a remarkable scholar in this custom, created contractualist speculations that stress the significance of supporting moral standards from the perspective of shared understanding.

Women's activist morals is another significant and developing region inside hypothetical morals that inspects the effect of orientation on moral hypothesis and practice. Women's activist ethicists scrutinize conventional moral hypotheses for their verifiable disregard of ladies' points of view and encounters. They underline the significance of resolving issues connected with orientation uniformity, care morals, and the convergences of orientation with other social classifications.

The morals of care, related with women's activist morals, centers around the ethical meaning of connections and the obligations that emerge from really focusing on others. Created by masterminds like Song Gilligan, care morals challenges the conventional accentuation on unique standards and features the significance of sympathy, empathy, and mindfulness of the necessities of others. This approach has been persuasive in reshaping conversations about moral training and the job of feelings in moral navigation.

Ecological morals is an arising field inside hypothetical morals that tends to the ethical elements of human communications with the climate. As worries about environmental change, biodiversity misfortune, and asset consumption heighten, ecological ethicists investigate the moral obligations people and social orders have toward the normal world. This viewpoint challenges anthropocentrism — the view that human interests outweigh those of different species — and advocates for a more comprehensive and environmentally reasonable ethic.

Strict and social points of view assume a critical part in molding hypothetical morals. Numerous moral speculations have establishes in strict practices, and strict morals keep on impacting moral talk. The Ten Charges in Judaism and Christianity, for instance, give a bunch of moral rules in view of heavenly power. Islamic morals correspondingly draws on the lessons of the Quran and the Hadith to direct moral lead. Eastern philosophical customs, for example, Confucianism and Buddhism, offer exceptional viewpoints on profound quality that contrast from Western methodologies.

Inside the domain of hypothetical morals, metaethics investigates the mystical and epistemological parts of profound quality. Savants in this field explore the idea of moral truth, the importance of moral language, and the reason for moral cases. Metaethical questions incorporate whether moral realities exist freely of human convictions, whether moral explanations express genuine insights, and how people come to be aware or find moral standards.

The connection among morals and human brain research is one more area of request inside hypothetical morals. Moral brain research investigates how people foster moral convictions, make moral decisions, and participate in moral way of behaving. This interdisciplinary field draws on experiences from brain science, neuroscience, and reasoning to comprehend the mental cycles and close to home factors that shape moral navigation. For instance, research in moral brain science plays analyzed the part of compassion, moral instincts, and the impact of social and social elements on moral turn of events.

Moral relativism is a metaethical position that declares the variety of moral convictions across societies and people. As indicated by moral relativism, there is no goal or all inclusive norm for deciding the rightness of moral decisions. All things being equal, moral standards are viewed as comparative with the social or individual setting in which they emerge. While moral relativism recognizes the significance of social variety and verifiable setting, it likewise raises difficulties connected with the potential for moral advancement and the capacity to investigate hurtful practices inside a specific culture.

The continuous discussions and advancements inside hypothetical morals feature the dynamic and developing nature of moral request. As new difficulties arise in regions like biotechnology, computerized reasoning, and worldwide equity, ethicists wrestle with the use of existing speculations to novel moral problems.

Issues encompassing the utilization of hereditary designing to change human characteristics, for example, bring up issues about the moral limits of mechanical mediations in human instinct and the possible ramifications for social balance.

The moral ramifications of arising innovations, like independent vehicles and high level observation frameworks, additionally brief moral reflection. Logicians and ethicists participate in conversations about the mindful turn of events and organization of these advances, taking into account issues connected with protection, independence, and the potential for unseen side-effects. The convergence of morals and innovation highlights the requirement for continuous moral request to address the difficulties of our quickly impacting world.

The job of pragmatic thinking in moral direction is a focal subject in hypothetical morals. Commonsense thinking includes the course of pondering and judgment through which people explore moral issues and settle on decisions that line up with their moral standards. Hypotheses of functional thinking investigate the mental cycles associated with weighing contending moral contemplations, surveying the importance of moral standards to explicit cases, and showing up at ethically supported choices.

Moral pondering frequently includes adjusting contending values and taking into account the viewpoints of various partners. The idea of moral pluralism recognizes the presence of different, possibly clashing, moral rules that people should explore. Moral pluralism features the intricacy of moral direction and underlines the significance of cautious reflection on the subtleties of specific circumstances. In tending to moral struggles, people might have to focus on specific qualities over others or look for a reasonable methodology that regards numerous ethical contemplations.

The advancement of moral person is one more key part of hypothetical morals. Prudence ethicists, specifically, underline the significance of developing highminded qualities that add to moral greatness. This course of character improvement includes the constant act of ethics like genuineness, respectability, fortitude, and sympathy. Moral schooling and moral development assume a significant part in molding people's characters and impacting their ability for moral way of behaving.

The topic of moral obligation is a repetitive subject in hypothetical morals. Speculations of moral obligation investigate the circumstances under which people are considered ethically responsible for their activities. Deciding moral obligation includes considering variables like goal, information, and the capacity to settle on independent decisions. Savants additionally look at the ramifications of moral obligation regarding issues like law enforcement, discipline, and cultural assumptions for responsibility.

The connection among morals and regulation is one more area of crossing point, with hypothetical morals giving an establishment to the turn of events and assessment of legitimate standards. Legitimate positivism, a way of thinking of regulation, declares that the authenticity of lawful principles isn't dependent upon their ethical rightness. Nonetheless, the association among regulation and morals stays a complex and discussed subject, with continuous conversations about the ethical legitimacy of specific regulations and the job of moral contemplations in legitimate direction.

Hypothetical morals additionally draws in with inquiries of civil rights, resolving issues of decency, uniformity, and the dissemination of assets inside social orders. Hypotheses of equity, for example, John Rawls' hypothesis of equity as reasonableness, propose standards for coordinating social foundations to guarantee an only dispersion of advantages and weights. Common agreement speculations, drawing on the thoughts of scholars like Hobbes, Locke, and Rousseau, investigate the theoretical arrangements that people would make to lay out an equitable and helpful society.

As well as inspecting moral standards at the individual and cultural levels, hypothetical morals stretches out its request to the worldwide field. Worldwide morals investigates the ethical components of worldwide relations, basic liberties, and worldwide administration. Savants in this field consider questions connected with the obligations of people, states, and worldwide associations in tending to worldwide difficulties, for example, neediness, environmental change, and equipped struggle. The idea of cosmopolitanism, which stresses a feeling of worldwide citizenship and shared honest convictions, has acquired conspicuousness in conversations about worldwide morals.

As hypothetical morals keeps on developing, contemporary savants draw in with interdisciplinary viewpoints and draw on experiences from fields like brain research, social science, financial matters, and political theory. This interdisciplinary methodology improves moral request by integrating exact examination and pragmatic contemplations into philosophical investigation. The cooperation among ethicists and experts in different fields mirrors a pledge to addressing certifiable moral difficulties and adding to the improvement of morally informed strategies and practices.

All in all, looking at the moral system inside hypothetical way of thinking gives a nuanced comprehension of the different points of view, hypotheses, and discussions that shape our way to deal with ethical quality. From deontological standards to consequentialist computations and the development of prudent person, moral speculations offer thorough systems for understanding and assessing human way of behaving. The continuous investigation of arising difficulties, moral ramifications of innovation, and worldwide moral contemplations highlights the powerful idea of hypothetical morals, forming how we might interpret profound quality and directing moral undertakings in a complex and consistently impacting world.

5.2 Discussing the principles of compassion, kindness, and interconnectedness.

Examining the standards of sympathy, graciousness, and interconnectedness includes investigating major parts of human profound quality and moral way of thinking. These standards, established in different social, strict, and philosophical customs, offer a focal point through which people and social orders can move toward moral navigation, relational connections, and the more extensive interconnectedness of every single living being.

Empathy, as a central moral standard, is much of the time characterized as the capacity to feel and comprehend the enduring of others and the craving to reduce that misery. This idea rises above social and strict limits and fills in as a foundation for moral conduct in various moral structures. Sympathy calls for compassion, an ability to resound with the encounters of others, encouraging a feeling of shared humankind inwardly. Thoughtfully, sympathy lines up with the morals of care, which accentuates the ethical meaning of connections and the obligations that emerge from really focusing on others.

The standard of graciousness supplements empathy and is described by kindness, liberality, and a certified worry for the prosperity of others. Generosity goes past simple good manners; it mirrors a more profound obligation to cultivating positive connections, supporting those out of luck, and adding to the making of a more amicable and sympathetic culture. Morally, thoughtfulness lines up with ethicalness morals, which stresses the advancement of temperate person characteristics.

Rehearsing consideration includes developing excellencies like liberality, persistence, and generosity, adding to the general moral prospering of people and networks.

Interconnectedness is a philosophical idea that highlights that every living being and peculiarities are interconnected and reliant. This standard rises above independence and features the interconnected idea of human life with the more extensive regular world. In Eastern ways of thinking, for example, Buddhism and Hinduism, interconnectedness is a focal precept, stressing the possibility that all creatures are important for a many-sided trap of presence. Morally, interconnectedness requires an acknowledgment of the effect of our activities on others and the climate, advancing a feeling of obligation for the prosperity of the whole environment.

The moral ramifications of empathy, benevolence, and interconnectedness reach out to different parts of human existence, including individual connections, cultural designs, and worldwide contemplations. In private connections, these standards guide people in exploring clashes, advancing comprehension, and encouraging a feeling of local area. Sympathy urges people to identify with the encounters of others, cultivating further associations and advancing a feeling of shared humankind.

Graciousness assumes a vital part in relational elements by establishing positive and steady conditions. Thoughtful gestures, whether little motions or more critical deeds, add to the prosperity of people and reinforce the social texture. The act of generosity in connections assists work with trusting, compassion, and a feeling of correspondence, cultivating a good and sustaining relational climate.

Interconnectedness, when applied to individual connections, features the far reaching influences of individual activities. Perceiving the interconnected idea of human life urges people to think about the more extensive outcomes of their conduct on others. This mindfulness advances a feeling of obligation and responsibility for the prosperity of those inside one's nearby circle and then some.

On a cultural level, the standards of sympathy, generosity, and interconnectedness have significant ramifications for the improvement of moral standards, social designs, and strategies. Empathy situated morals challenge cultural designs that propagate imbalance, separation, and bad form. The morals of care, which puts significance on connections and common reliance, calls for social arrangements that focus on the prosperity of weak populaces and address foundational issues that add to misery.

Benevolence, when applied to cultural designs, requires the making of comprehensive and strong networks. Strategies that advance social government assistance, training, and medical care add to a kinder society that esteems the poise and prospering of every one of its individuals. The moral standards of graciousness likewise challenge cultural standards that propagate savagery, double-dealing, or detachment to the enduring of others.

Interconnectedness, when applied to cultural morals, underscores the significance of perceiving the worldwide effect of nearby activities. Ecological morals, for example, highlights the interconnected connection between human exercises and the strength of the planet. Strategies that focus on supportability, protection, and dependable asset the board mirror a moral obligation to the interconnectedness of every single living being and the regular world.

Worldwide moral contemplations additionally become an integral factor while examining empathy, benevolence, and interconnectedness. In a world that is progressively interconnected through innovation, exchange, and correspondence, moral standards should reach out past individual and cultural limits. Sympathy on a worldwide scale includes resolving issues like neediness, struggle, and helpful emergencies with a guarantee to reducing enduring and advancing equity.

Consideration, when expanded worldwide, calls for global collaboration, strategy, and endeavors to resolve foundational issues that add to disparity and foul play. Thoughtful gestures at a worldwide level might include giving guide to networks impacted by catastrophic events, supporting for basic liberties, or adding to drives that advance worldwide prosperity.

Interconnectedness, when applied universally, underscores the common obligation regarding tending to squeezing difficulties that influence humankind in general. Environmental change, for instance, requires cooperative endeavors from countries all over the planet to relieve its effect and adjust to the changing ecological circumstances. Perceiving the interconnected idea of worldwide difficulties prompts moral contemplations that focus on collaboration, fortitude, and the prosperity of people in the future.

Rationally, conversations about sympathy, generosity, and interconnectedness meet with different moral speculations. Ideals morals, which stresses the advancement of idealistic person characteristics, adjusts intimately with the standards of sympathy and thoughtfulness. Temperate people, as per righteousness morals, are the individuals who normally epitomize characteristics like sympathy, kindheartedness, and a feeling of interconnectedness.

Deontological morals, with its accentuation on moral obligations and standards, can give a system to understanding the moral commitments related with sympathy and generosity. The acknowledgment of the inborn worth and pride of each and every person, a focal fundamental in deontology, lines up with the standards of empathy and generosity.

Consequentialist morals, especially utilitarianism, can evaluate the virtue of activities in light of their general effect on prosperity. Demonstrations of sympathy and thoughtfulness, which add to the joy and prospering of people and networks, might be viewed as ethically laudable according to a consequentialist viewpoint.

The morals of care, as a women's activist moral viewpoint, places sympathy, generosity, and interconnectedness at the focal point of moral contemplations. Care morals challenges customary thoughts of withdrawn, unbiased thinking and accentuates the significance of connections, sympathy, and the obligations that emerge from really focusing on others.

Strict and otherworldly customs likewise assume a critical part in forming conversations about sympathy, generosity, and interconnectedness. Numerous strict lessons underscore the significance of affection, empathy, and benevolence as focal precepts of moral living. For instance, the Christian rule of agape love urges devotees to stretch out sympathy and graciousness to all, mirroring a feeling of interconnectedness established in a common heavenly creation.

Buddhist lessons on empathy (karuna) and interconnectedness (pratitya-samutpada) feature the moral basic of perceiving the relationship of every living being and developing a humane disposition toward others' torment. Hindu practices comparably underline the interconnectedness of all life through the idea of dharma, which envelops moral obligations and obligations.

All in all, the standards of sympathy, graciousness, and interconnectedness structure a rich moral embroidery that rises above social, strict, and philosophical limits. These standards guide people in exploring individual connections, illuminate cultural designs and strategies, and brief moral contemplations on a worldwide scale. Thoughtfully, these standards cross with different moral speculations, from ethicalness morals to deontology and consequentialism, offering a comprehensive system for understanding and exploring the intricacies of human profound quality. The continuous investigation and utilization of these standards add to the making of a more caring, kind, and interconnected world.

5.3 Exploring how Theorophical ethics can guide individuals in their daily lives.

Investigating how hypothetical morals can direct people in their day to day routines includes looking at how moral hypotheses and standards give a system to moral independent direction, self-awareness, and exploring the intricacies of human connections. Hypothetical morals, enveloping different moral viewpoints, for example, deontology, consequentialism, ideals morals, and others, offers bits of knowledge into what is viewed as ethically right or wrong, giving people apparatuses to evaluate and address moral quandaries in their everyday encounters.

One manner by which hypothetical morals guides people in their day to day routines is through the standards of deontology. Deontological morals, got from the Greek word "deon" meaning obligation, sets that people have objective moral obligations they are committed to follow, no matter what the results. Immanuel Kant, a focal figure in deontological morals, contended that ethical activities ought to be directed by widespread rules that can be applied reliably across various circumstances.

In commonsense terms, deontological morals guides people by underlining the significance of complying with moral guidelines and standards. For instance, the rule of trustworthiness as an ethical obligation implies that people are urged to come clean and maintain the trustworthy worth, in any event, while confronting testing conditions. This can direct people in their day to day collaborations, elevating a guarantee to honesty and moral conduct in different individual and expert settings.

Also, deontological standards frequently incorporate the regard for the inborn nobility of people. This guides people to approach others with deference, perceiving their independence and worth. In day to day existence, this guideline might appear through activities, for example, regarding others' protection, respecting responsibilities, and shunning activities that abuse the freedoms or prosperity of others.

Consequentialist morals, one more major moral hypothesis, guides people by zeroing in on the results or outcomes of their activities. Utilitarianism, a type of consequentialism, sets that the ethical quality of a still up in the air by the general satisfaction or joy it produces.

This point of view urges people to consider the results of their activities on the prosperity of themselves as well as other people.

In commonsense terms, utilitarianism guides people to go with choices that add to the best by and large bliss. This might include considering the possible effect of one's activities on the joy and prosperity of others, weighing contending interests, and pursuing decisions that intend to augment generally speaking government assistance. For instance, people could participate in thoughtful gestures, liberality, or local area administration as a feature of their day to day routines, directed by the consequentialist guideline of advancing bliss and limiting misery.

Ethicalness morals, established in antiquated Greek way of thinking, guides people by accentuating the advancement of upright person qualities. Aristotle, a vital figure in uprightness morals, contended that ethical way of behaving emerges from developing temperances like boldness, genuineness, and empathy. Ethicalness morals goes past endorsing explicit principles and on second thought centers around the sort of individual one ought to turn into.

In day to day existence, excellence morals guides people to participate in rehearses that develop ethical person qualities. This includes going with deliberate decisions to foster characteristics like compassion, lowliness, and versatility. For instance, an individual rehearsing ideals morals may effectively look for chances to show sympathy, practice genuineness, and exhibit boldness notwithstanding challenges, in this manner epitomizing idealistic characteristics in their everyday connections.

The morals of care, a methodology inside ideals morals, further aides people in their regular routines by underlining the ethical meaning of connections and the obligations that emerge from really focusing on others. Care morals challenges conventional moral speculations by featuring the significance of sympathy, empathy, and mindfulness of the necessities of others.

In reasonable terms, care morals guides people to focus on connections and providing care liabilities. This might include thoughtful gestures, daily reassurance, and sustaining ways of behaving in private connections, families, and networks. For example, a parental figure rehearsing care morals could focus on the prosperity of those under their consideration, settling on choices that mirror a promise to cultivating association, understanding, and responsiveness to the necessities of others.

The standards of sympathy, consideration, and interconnectedness, which rise above unambiguous moral hypotheses, additionally guide people in their day to day routines. Sympathy includes the capacity to feel and comprehend the enduring of others and the longing to lighten that affliction. In everyday connections, people directed by sympathy might show compassion, offer help, and take part in thoughtful gestures that add to the prosperity of others.

Consideration, as a guideline, urges people to develop kindness, liberality, and positive communications. In day to day existence, rehearsing thoughtfulness might include little signals like keeping the door open for somebody, offering a caring word, or effectively paying attention to other people. These thoughtful gestures add to establishing a positive and strong social climate.

Interconnectedness, as a core value, prompts people to perceive their relationship with others and the regular world. This mindfulness energizes dependable and moral decisions that think about the more extensive effect of individual activities. For example, people directed by a feeling of interconnectedness might settle on earth cognizant choices, add to social causes, and take part in endeavors to address worldwide difficulties.

Hypothetical morals likewise directs people in their self-awareness by giving a system to reflection and self-assessment. Moral speculations urge people to investigate their qualities, standards, and convictions, cultivating a more profound comprehension of their own ethical compass. This mindfulness is essential for settling on moral choices lined up with one's qualities and for persistently refining and developing one's moral standpoint.

Moreover, hypothetical morals guides people in exploring moral predicaments by offering an efficient way to deal with moral thinking. When confronted with complex moral choices, people can go to moral speculations to break down the circumstance, weigh contending values, and think about the possible results of various blueprints. This course of moral consideration empowers people to settle on informed and ethically legitimate choices in different parts of their lives.

Logically, hypothetical morals prompts people to participate in continuous moral request. By wrestling with major inquiries regarding profound quality, people can extend how they might interpret moral standards, challenge suspicions, and foster a more nuanced and intelligible moral perspective. This scholarly commitment adds to self-awareness, moral development, and the capacity to explore the intricacies of moral dynamic in a steadily impacting world.

In the domain of individual connections, hypothetical morals guides people in cultivating sound and moral associations with others. The standards of regard, empathy, and graciousness act as starting points for building significant connections. People directed by moral standards are bound to take part in open correspondence, common comprehension, and cooperative critical thinking inside their own connections.

For instance, in a close connection, moral contemplations might include regarding the independence and prosperity of one's accomplice, rehearsing trustworthiness and straightforwardness, and tending to clashes in a valuable and sympathetic way. Essentially, in companionships and familial connections, moral standards guide people to focus on trust, support, and the in general thriving of those they care about.

The use of hypothetical morals in the expert circle is similarly huge. Moral contemplations guide people in settling on choices that line up with their qualities and add to the prosperity of others. Experts, whether in business, medical services, schooling, or different fields, benefit from a strong moral establishment that illuminates their activities and navigation.

Deontological standards, for example, trustworthiness and respectability, guide experts to maintain moral norms in their work. For example, in business, people directed by deontological morals might focus on genuineness in exchanges, satisfy authoritative commitments, and consider the more extensive cultural effect of strategic approaches. In medical services, deontological standards might appear through a promise to patient independence, privacy, and the arrangement of value care.

Consequentialist contemplations, then again, guide experts to survey the likely results of their activities. For example, in business, consequentialist morals might include assessing the social and ecological effect of business choices. In medical care, consequentialist contemplations might illuminate choices about therapy plans, considering the general prosperity and satisfaction of patients.

Ideals morals, with its accentuation on character improvement, is especially important in the expert domain. Highminded experts are the people who typify characteristics like respectability, decency, and sympathy. In a corporate setting, upright pioneers might focus on moral direction, decency in worker relations, and a promise to social obligation.

Chapter 6

Unity of All Religions

In the immense embroidery of human life, woven with the strings of different societies, convictions, and customs, the idea of solidarity among religions remains as an encouraging sign and understanding. Since forever ago, mankind has wrestled with inquiries of presence, reason, and the heavenly, looking for comfort and insight through different profound ways. Notwithstanding the evident contrasts that recognize one confidence from another, a nearer assessment uncovers fundamental shared traits that propose a common quintessence among the world's religions.

At the core of numerous strict practices lies a crucial affirmation of a higher power or extraordinary reality. Whether communicated as God, divine beings, the Outright, or by different names, the acknowledgment of a power past the substantial and quantifiable joins innumerable devotees across the globe. This consistent idea highlights the significant human longing for association with an option that could be more noteworthy than oneself, a mission that rises above social and topographical limits.

One can follow the foundations of strict solidarity to the beginning of human advancement, where old social orders created multifaceted conviction frameworks to figure out their general surroundings. The animistic convictions of native societies, the polytheistic pantheons of old civilizations like Greece and Rome, and the monotheistic disclosures of prophets and saviors all offer a typical reason - to give a structure to figuring out the secrets of presence and exploring the intricacies of human existence.

While the overt gestures of confidence might contrast, the inward encounters of stunningness, worship, and the quest for the heavenly remain strikingly comparative. Otherworldly experiences, pondering practices, and ceremonies that lift the human soul are tracked down in fluctuating structures across strict customs. The Sufi spinning dervishes look for association with the heavenly through euphoric dance, reflecting the Christian spiritualists who portray the "blissful vision" and the Hindu yogis as they continued looking for Samadhi.

The moral lessons implanted in strict regulations additionally highlight shared values that advance sympathy, equity, and moral direct. The Brilliant Rule, a guideline present in different structures in Christianity, Judaism, Islam, Buddhism, and others, epitomizes the pith of regarding others as one might want to be dealt with. Past the doctrinal incongruities, moral goals guide devotees to carry on with idealistic existences, cultivating a feeling of obligation toward each other and the world.

Regardless of these common components, history likewise demonstrates the veracity of the struggles and divisions that have emerged for the sake of religion. The Campaigns, strict conflicts, and partisan difficulty highlight the difficulties intrinsic in exploring the variety of beliefs. Notwithstanding, advocates of interfaith exchange and ecumenism contend that these contentions frequently come from mistaken assumptions and misinterpretations instead of innate beyond reconciliation contrasts.

In the advanced time, the call for strict solidarity has picked up speed as worldwide network and intercultural trade become progressively pervasive. Interfaith drives try to encourage shared understanding and regard among devotees of various religions, underlining shared belief while recognizing and valuing contrasts. The Parliament of the World's Religions, established in 1893, remains as a demonstration of the continuous endeavors to advance discourse and participation among the world's strict networks.

Moreover, the universalization of basic freedoms has given a typical moral system that rises above strict limits. Ideas like opportunity of heart, equity, and poise resound across societies, provoking social orders to maintain these standards paying little mind to strict connection. The quest for civil rights turns into a common undertaking as people of different religions join to resolve issues of neediness, imbalance, and ecological debasement.

In investigating the solidarity of religions, perceiving the intrinsic variety inside every tradition is fundamental. The branches and groups inside Christianity, the factions in Islam, the ways of thinking in Hinduism, and the different understandings inside Buddhism embody the rich embroidery of points of view that exist even inside a solitary confidence. Embracing this variety becomes basic to cultivating solidarity, as it recognizes the assortment of human encounters and articulations of the sacrosanct.

The idea of strict pluralism, which declares that various religions are substantial ways to the heavenly, challenges exclusivist sees that guarantee one confidence as the sole store of truth. Defenders of pluralism contend that perceiving the legitimacy of assorted profound ways doesn't reduce the meaning of individual confidence customs but instead upgrades the aggregate embroidered artwork of human otherworldliness.

In the journey for strict solidarity, the job of discourse becomes fundamental. Veritable discourse requires a receptiveness to tuning in, understanding, and valuing the viewpoints of others. It includes rising above assumptions and generalizations, considering the acknowledgment of shared desires and values. Interfaith exchange looks to fabricate scaffolds of understanding, encouraging a feeling of collaboration that rises above strict partitions.

Scholars and researchers took part in relative strict examinations assume a urgent part in explaining the shared characteristics and contrasts among world religions. By inspecting holy texts, ceremonies, and philosophical ideas, these researchers add to a more profound comprehension of the all inclusive subjects that support strict practices. Relative religion features shared moral standards as well as uncovers the social and authentic settings that shape different articulations of confidence.

Inside the domain of otherworldliness, the idea of the perpetual way of thinking places a widespread center of magical and mystical bits of insight that rises above the surface varieties of strict practices. Aldous Huxley, in his investigation of the enduring way of thinking, recommends that underneath the different social structures lies an immortal and general insight that has been explained by spiritualists, sages, and scholars all through the ages.

Otherworldliness, with its accentuation on immediate, individual experience of the heavenly, frequently fills in as a gathering point for searchers across various religions. The enchanted customs inside Christianity, for example, the works of Meister Eckhart and the Haze of Accidental, find reverberations in the Sufi supernatural quality of Islam, the Kabbalistic practices in Judaism, and the pensive practices in Hinduism and Buddhism. The unitive encounters depicted by spiritualists share a consistent idea, rising above the limits of strict marks.

In the domain of reasoning, the enduring way of thinking finds articulation in progress of researchers like Huston Smith, who devoted his life to investigating the world's religions. Smith's original work, "The World's Religions," digs into the center standards of significant strict customs, underlining the all inclusiveness of profound bits of knowledge while regarding the variety of social articulations.

The acknowledgment of a general profound embodiment doesn't discredit the meaning of social and verifiable settings in molding strict practices. Every confidence conveys a novel social engraving, enhanced by the narratives, images, and customs that have developed over hundreds of years. The festival of variety inside solidarity recognizes the magnificence of this embroidered artwork, where each string adds to the wealth of the entirety.

Craftsmanship and writing give one more road to investigating the solidarity of religions. Over the entire course of time, craftsmen and journalists have drawn motivation from different strict customs, making works that resound across social limits. The verse of Rumi, drawing from Islamic enchantment, addresses searchers, everything being equal. The canvases of William Blake, implanted with otherworldly vision, rise above the limits of a particular strict custom.

The idea of the aggregate oblivious, as proposed via Carl Jung, recommends that mankind shares a supply of prototype images and pictures that track down articulation in legends, dreams, and strict stories. Jung's investigation of the aggregate oblivious focuses to a common mental and profound legacy that rises above the limits of individual societies and religions.

In the domain of fantasy and imagery, Joseph Campbell's relative folklore investigates the normal topics and themes that repeat in the legends of assorted societies. The legend's excursion, the grandiose battle among great and wickedness, and the journey for illumination arise as widespread accounts that reverberate across strict and social limits. Campbell's work welcomes us to look past the surface distinctions of strict stories and perceive the basic model examples that interface them.

As the world turns out to be more interconnected, the requirement for strict education and intercultural understanding turns out to be progressively indispensable. Training assumes a key part in encouraging appreciation for different conviction frameworks and advancing a nuanced comprehension of the intricacies of strict personality. By integrating the investigation of world religions into instructive educational programs, social orders can develop a worldwide viewpoint that rises above restricted sectarianism.

The job of strict forerunners in advancing solidarity couldn't possibly be more significant. Pioneers from different religions have an obligation to show resistance, empathy, and collaboration. Interfaith coordinated efforts on philanthropic tasks, civil rights drives, and natural stewardship give functional roads to strict pioneers to show shared values in real life.

The difficulties to strict solidarity are, notwithstanding, complex. Political, social, and financial factors frequently interweave with strict character, prompting pressures and clashes. Verifiable complaints, international contemplations, and power elements can muddle endeavors to cultivate understanding and coordinated effort among strict networks. Tending to these difficulties requires an all encompassing methodology that thinks about the interconnectedness of strict, social, and political domains.

The job of ladies in religion likewise justifies consideration in conversations of solidarity. Numerous strict customs have wrestled with issues of orientation fairness, and the incorporation of different voices is critical for a more complete comprehension of otherworldliness. Endeavors to advance orientation value inside strict organizations add to a more comprehensive and delegate articulation of confidence.

Ecological worries give a shared conviction to strict networks to team up. The interconnectedness of all life, a topic present in numerous profound customs, lines up with the developing consciousness of environmental reliance. The common obligation regarding natural stewardship rises above strict contrasts, inciting devotees to cooperate for the prosperity of the planet.

The development of a worldwide ethic, as expressed by Hans Küng, imagines an essential arrangement of moral rules that can be embraced by people of all strict and philosophical foundations. Küng's vision underscores the normal qualities that support different strict customs, giving a premise to common regard and cooperation. A worldwide ethic turns into a binding together power that rises above strict particularities while regarding the rich variety of human otherworldliness.

All in all, the solidarity of all religions arises as a significant and optimistic vision for humankind. Notwithstanding the surface qualifications that portray one confidence from another, the basic flows of otherworldliness, morals, and the mission for the heavenly weave an embroidery that interfaces the world's religions. Perceiving this solidarity requires a readiness to participate in exchange, appreciate variety, and embrace the common qualities that tight spot humankind together. As the world keeps on developing, the quest for strict solidarity remains as an encouraging sign, welcoming people and networks to rise above the limits that different and to find the shared belief that joins together.

6.1 Investigating the concept of the unity of all religions in Theorophy.

In digging into the investigation of the solidarity, everything being equal, the discipline of Theorophy arises as a focal point through which to look at the interconnections and shared rules that support different strict customs. Theorophy, a term begat to typify the blend of philosophy and reasoning, looks to disentangle the all inclusive insights that navigate strict limits. Established in the conviction that there exists a binding together string winding through the texture of all religions, Theorophy welcomes investigation into the pith of otherworldliness, the idea of the heavenly, and the common moral objectives that guide human lead.

At its center, Theorophy sets that underneath the doctrinal contrasts and social articulations lies a typical wellspring of intelligence. This viewpoint rises above the limits of a specific strict system, welcoming followers of different religions to take part in an exchange that moves past the superficial qualifications. Theorophy, as an inter-disciplinary field, draws upon philosophical bits of knowledge, philosophical thinking, and similar investigations of strict practices to recognize the widespread rules that tight spot mankind in its profound journey.

The mission for the solidarity of all religions inside Theorophy starts with an investigation of the idea of the heavenly. Different strict practices conceptualize the heavenly in different ways - as an individual god, a generic power, or a comprehensive enormous standard. Theorophy places that underneath these different articulations lies an extraordinary reality that evades exact definition yet fills in as the source and sustainer of all that exists.

In analyzing philosophical precepts, Theorophy looks to distinguish the normal subjects that reverberate across strict practices. The idea of a Maker, a sustainer of the universe, and a generous power that underlies presence shows up in mono-theistic practices like Judaism, Christianity, and Islam. Essentially, the Hindu idea of Brahman as a definitive reality and the Buddhist comprehension of the indistinct, un-conditioned Nirvana mirror a reverberation with the possibility of an extraordinary, unspeakable heavenly rule.

Theorophy recognizes that the variety of strict articulations emerges from social, verifiable, and phonetic settings. While the religious language and emblematic portrayals might shift, the hidden affirmation of a heavenly presence stays a common part of human otherworldliness. Theorophical request welcomes adherents to rise above the restrictions of sectarianism and perceive the shared conviction that exists in how they might interpret the heavenly.

Morals, a focal part of strict lessons, turns into one more point of convergence in the investigation of Theorophy. The moral standards upheld by various religions frequently combine on key ideas like sympathy, equity, and the Brilliant Rule. The call to regard others as one might want to be dealt with tracks down articulation in Christianity, Judaism, Islam, Buddhism, and different other confidence customs. Theorophy fights that these common moral objectives mirror a general moral compass that rises above strict limits, underscoring the intrinsic pride and worth of each and every person.

Moreover, Theorophy digs into the domain of supernatural encounters that penetrate strict customs. Enchantment, with its accentuation on direct fellowship with the heavenly, offers an exceptional vantage point for investigating the solidarity of religions. The spiritualist's experience with the indescribable, whether communicated through Christian pensive practices, Sufi spinning, or Eastern thoughtful customs, focuses to a common human ability to rise above the customary and interface with an otherworldly reality.

Theorophy draws motivation from the lasting way of thinking, an idea that recommends the presence of a widespread center of enchanted and powerful insights. This viewpoint, supported by researchers like Aldous Huxley, sets that underneath the surface variety of strict practices lies an immortal and all inclusive insight. Theorophy, in arrangement with the lasting way of thinking, perceives that spiritualists, sages, and scholars over the entire course of time have verbalized bits of knowledge that reverberation across social and strict limits.

Similar religion, a foundation of Theorophy, includes a thorough assessment of sacrosanct texts, customs, and religious ideas across various beliefs. Researchers participated in near strict examinations add to the explanation of shared traits and qualifications, encouraging a more profound comprehension of the general subjects that support different strict customs. Theorophy, by integrating these similar bits of knowledge, tries to uncover the common profound legacy that joins mankind.

The investigation of Theorophy stretches out past the scholastic domain, including the functional element of interfaith discourse. Interfaith drives, directed by the standards of Theorophy, mean to work with understanding and cooperation among people of various strict foundations. The Parliament of the World's Religions, established in 1893, fills in as a worldwide stage for encouraging interfaith discourse and participation, exemplifying the soul of Theorophy in real life.

In the contemporary scene, where globalization and mechanical interconnectedness bring assorted societies into closer contact, the significance of Theorophy turns out to be progressively obvious. The difficulties of strict pluralism, social variety, and the requirement for common comprehension require a structure that rises above restricted sectarianism. Theorophy offers a point of view that urges people to look past the surface differentiations of strict customs and perceive the common qualities and desires that tight spot mankind together.

The job of Theorophy in tending to strict struggles is especially critical. Many struggles since forever ago and in the current day have emerged from errors, misinterpretations, and dug in doctrines. Theorophy, by advancing a nuanced and compassionate comprehension of different strict viewpoints, adds to the goal of contentions and the development of a more agreeable conjunction.

A vital part of Theorophy is the acknowledgment of the variety inside solidarity. While it stresses the widespread rules that join religions, it additionally recognizes the wealth of individual social and strict articulations. The branches and groups inside every confidence custom, the social subtleties in ceremonies and practices, and the authentic settings that shape strict stories are fundamental parts of the perplexing embroidery of human otherworldliness.

Theorophy stretches out its request to the domain of reasoning, investigating the mystical and epistemological establishments that support strict convictions. The connection point among reason and confidence, the idea of disclosure, and the manners by which people capture the heavenly become subjects of consideration inside Theorophical talk. This interdisciplinary methodology looks to coordinate the bits of knowledge of reasoning with the philosophical viewpoints present in different strict customs.

Theorophy likewise perceives the significance of individual profound encounters in the mission for strict solidarity. Individual experiences with the heavenly, extraordinary snapshots of understanding, and the inward excursion of self-revelation all add to the more extensive woven artwork of human otherworldliness. Theorophy, by esteeming and recognizing the variety of individual profound ways, empowers a more comprehensive and all encompassing comprehension of strict solidarity.

The difficulties looked by Theorophy in its quest for the solidarity of all religions are complex. Suspicion and obstruction might emerge from dug in doctrinal positions, social biases, and the feeling of dread toward weakening the uniqueness of individual confidence customs. Theorophy, in any case, declares that perceiving the common quintessence of religions doesn't lessen the peculiarity of every practice yet rather improves the aggregate comprehension of otherworldliness.

In tending to the intricacies of strict solidarity, Theorophy battles with the sociopolitical aspects that frequently meet with strict personality. Authentic complaints, power elements, and international contemplations can convolute the endeavors to encourage understanding and cooperation among strict networks. Theorophy recognizes the requirement for an all encompassing methodology that thinks about the interconnectedness of strict, social, and political domains.

The job of schooling arises as an extraordinary power inside Theorophy. By integrating the investigation of Theorophy into instructive educational plans, social orders can support a worldwide viewpoint that rises above limited sectarianism. Strict proficiency, social getting it, and the capacity to take part in aware discourse become fundamental parts of schooling that encourages the solidarity, all things considered.

Strict pioneers, as persuasive figures inside their networks, assume a vital part in advancing strict solidarity. Theorophy calls upon strict pioneers to display resistance, sympathy, and collaboration. Interfaith joint efforts on compassionate undertakings, civil rights drives, and ecological stewardship give unmistakable articulations of shared values in real life.

The consideration of ladies in conversations of Theorophy becomes basic for a more far reaching comprehension of strict solidarity. Numerous strict practices have wrestled with issues of orientation uniformity, and the acknowledgment of different voices adds to a more comprehensive and delegate articulation of confidence. Theorophy, in embracing orientation value, recognizes the significance of a variety of points of view in the investigation of otherworldly insights.

Ecological worries, with their worldwide ramifications, act as a shared conviction for strict networks to team up. The interconnectedness of all life, a topic present in numerous profound practices, lines up with the developing consciousness of environmental.

6.2 Highlighting common threads among various religious traditions.

In the huge woven artwork of human otherworldliness, different strict customs arise as unmistakable strings, each winding around its special account of the holy. In the midst of the evident contrasts in ceremonies, regulations, and social articulations, a nearer assessment uncovers an unpredictable example of consistent ideas that join different religions. This investigation into the common components among various religions enlightens the hidden widespread rules that rise above the limits of time, topography, and doctrinal particulars.

Acknowledgment of the Otherworldly:

At the core of numerous strict customs lies the affirmation of an otherworldly reality or heavenly presence. Whether communicated as a solitary, individual God, a variety of divinities, or a sweeping inestimable power, the acknowledgment of a more powerful fills in as a basic component. This consistent idea joins monotheistic practices like Christianity, Judaism, and Islam, as well as polytheistic conviction frameworks like Hinduism and old legends. The longing to interface with the extraordinary highlights a common human mission for importance and reason past the material domain.

Moral Objectives:

Installed inside the lessons of different religions are moral rules that guide human lead. The Brilliant Rule, epitomized in regarding others as one might want to be dealt with, arises as a widespread moral objective. This guideline reverberations across strict practices, rising above social and doctrinal contrasts. Whether verbalized in the expressions of Jesus in Christianity, the idioms of Prophet Muhammad in Islam, or as a crucial idea in Buddhism and Hinduism, the call for sympathy, equity, and moral direct fills in as a typical moral establishment.

Customs and Holy Practices:

While the particular customs and practices might change broadly, the drive to participate in emblematic demonstrations of love and association with the heavenly is a common trademark. Ceremonies, like supplication, reflection, reciting, and public love, structure fundamental parts of strict articulation across customs. The construction and frame might vary, however the hidden aim to develop a feeling of holiness, express dedication, and lay out an association with the heavenly stay steady.

Hallowed Texts and Intelligence Writing:

Strict practices frequently have sacrosanct texts or sacred writings that act as archives of insight, direction, and heavenly disclosure. Whether it be the Holy book in Christianity, the Quran in Islam, the Vedas in Hinduism, or the Tao Te Ching in Taoism, these texts offer moral lessons, accounts of beginning, and experiences into the idea of presence. The veneration for sacrosanct writing as a wellspring of heavenly correspondence is a common component, mirroring a shared trait chasing otherworldly insights.

The Idea of Penance:

The subject of penance, both emblematic and exacting, penetrates different strict accounts. Whether depicted through ceremonial contributions, demonstrations of magnanimity, or the idea of vicarious expiation, penance implies a profound obligation to the heavenly and the more extensive local area. The imagery of penance, exemplified in customs like Christianity with the execution of Jesus or Hinduism with Yajna services, highlights the general theme of giving up individual cravings for a higher motivation.

Magical and Insightful Practices:

Inside numerous strict customs, there exists a mysterious aspect that underlines direct fellowship with the heavenly. Spiritualists, contemplatives, and religious zealots across societies look for significant, extraordinary encounters that rise above the conventional limits of discernment. Whether it be the Christian spiritualists investigating the "dim evening of the spirit," the Sufi custom of euphoric spinning, or the reflective acts of Buddhist priests, the quest for an immediate, individual experience with the heavenly uncovers a common human limit with regards to otherworldly experience.

Imagery and Models:

Emblematic language and original symbolism plague strict practices, offering a rich embroidery of representations and moral stories. Normal originals, like the legend's excursion, the vast battle among great and wickedness, and the theme of revival, repeat in fantasies, legends, and strict accounts across societies. These widespread images rise above etymological and social boundaries, addressing central parts of the human mind and the aggregate oblivious.

Cosmology and Creation Fantasies:

Investigations of the beginning and nature of the universe highlight conspicuously in strict cosmologies. Creation legends, whether relating a heavenly demonstration of creation, an enormous egg, or the rising up out of early stage mayhem, share topical similitudes. The cosmogonic stories in strict practices, from the Beginning record in Judaism and Christianity to the cosmogonies in Hinduism and native folklores, mirror a typical human interest in the starting points of the universe and mankind's place inside it.

The Idea of Heavenly Love:

The thought of heavenly love, enveloping sympathy, kindness, and generosity, arises as a focal topic in numerous strict customs. Whether communicated as agape in Christianity, rahmah in Islam, or karuna in Buddhism, the heavenly trait of adoration rises above social and strict limits. Developing adoration for the heavenly and stretching out that adoration to individual creatures is a binding together string that highlights the interconnectedness of mankind.

Patterns of Life, Demise, and Restoration:

The repetitive idea of presence, set apart by topics of life, passing, and reestablishment, tracks down articulation in strict rituals, celebrations, and philosophical viewpoints. The pattern of seasons, the life-demise resurrection theme, and the acknowledgment of temporariness are repeating topics. Whether celebrated through the Christian story of revival, the Hindu idea of samsara, or the imagery of death and resurrection in native practices, the recurrent idea of life turns into a common thought.

Interconnectedness and Reliance:

Numerous strict customs articulate a perspective that underlines the interconnectedness and relationship of all life. This environmental mindfulness, established in profound lessons, lines up with contemporary natural worries. The acknowledgment of a common obligation regarding the prosperity of the planet is clear in the stewardship ethos present in native insight, the respect for nature in Shintoism, and the interconnectedness underscored in the lessons of Francis of Assisi in Christianity.

The Quest for Internal Change:

Across different strict customs, a typical accentuation on inward change and profound development is clear. Practices like self-control, self-assessment, and the development of excellencies target rising above egoic restrictions and arousing to a higher condition of cognizance. Whether explained through the Christian idea of purification, the Islamic idea of tazkiyah, or the Buddhist way of edification, the quest for inward change mirrors a common yearning for individual and profound turn of events.

Trust and Reclamation:

The topic of expectation and the mission for reclamation notwithstanding difficulty reverberate across strict accounts. Whether communicated through the Christian confidence in salvation, the Islamic idea of contrition and pardoning, or the Hindu thought of freedom (moksha), the human longing for greatness and the chance of change notwithstanding enduring and blemish is a common theme.

All inclusive Moral Regulation:

The acknowledgment of a widespread moral regulation, frequently reflected in the arrangement of human regulations with divine standards, is a repetitive topic in strict practices. The possibility that moral standards are established in an extraordinary source, whether divine edicts or regular regulation, highlights a common perspective of the need for moral rules to direct human way of behaving.

In total, the investigation of ongoing ideas among different strict practices uncovers a rich embroidery of shared standards, subjects, and desires. In spite of the superficial qualifications and doctrinal contrasts, the comprehensiveness of human encounters and the mission for the sacrosanct become apparent. This acknowledgment of shared view welcomes people to rise above sectarianism, cultivating common grasping, regard, and participation across assorted strict scenes. The common qualities implanted in these consistent ideas act as an establishment for exchange, interfaith joint effort, and the aggregate quest for an additional agreeable and interconnected world.

6.3 Discussing how Theorophy promotes interfaith understanding and harmony.

Theorophy, as an all encompassing and interdisciplinary way to deal with investigating the solidarity of all religions, assumes a significant part in advancing interfaith comprehension and concordance. Established in the combination of philosophy and reasoning, Theorophy welcomes people to dig into the consistent ideas that go through assorted strict practices while valuing the remarkable social and verifiable articulations that shape every confidence. In looking at how Theorophy cultivates interfaith comprehension, a few key viewpoints come to the very front.

1. Acknowledgment of Shared Profound Standards:

   At the center of Theorophy is the acknowledgment of divided profound rules that rise above the superficial qualifications between religions. The investigation of all inclusive topics like the idea of the heavenly, moral objectives, and the quest for inward change gives a shared belief to devotees of various religions. By stressing these common standards, Theorophy spans holes and encourages an appreciation for the interconnectedness of different otherworldly ways.

2. Relative Strict Examinations:

   Theorophy puts areas of strength for an on similar strict investigations, empowering researchers and experts to participate in a thorough assessment of hallowed texts, ceremonies, and philosophical ideas across various confidence customs. This academic undertaking reveals insight into the shared characteristics that support strict articulations, uncovering the general experiences that rise up out of different social and authentic settings. Near examinations add to a more profound comprehension of the common otherworldly legacy that joins mankind.

3. Exchange and Correspondence:

   A crucial part of Theorophy is the advancement of exchange and correspondence among people of various strict foundations. Interfaith exchange, directed by the standards of Theorophy, energizes open and aware discussions that rise above doctrinal contrasts. Through significant trades, members gain bits of knowledge into the convictions, practices, and profound encounters of others, encouraging common getting it and dispersing confusions.

4. Embracing Strict Variety:

   Theorophy recognizes the rich embroidered artwork of strict variety, perceiving that every confidence custom contributes interesting viewpoints to the aggregate comprehension of otherworldliness. Instead of looking to homogenize strict articulations, Theorophy commends the variety inside solidarity. This comprehensive methodology urges people to see the value in the social subtleties, customs, and religious subtleties that describe various religions.

5. The Lasting Way of thinking:

The idea of the perpetual way of thinking, a focal subject in Theorophy, declares the presence of a widespread center of otherworldly and supernatural bits of insight that rises above the surface varieties of strict customs. Theorophy draws upon the experiences of spiritualists, sages, and scholars since forever ago who have enunciated an immortal and all inclusive insight. By underlining these lasting insights, Theorophy advances a mutual perspective that rises above strict limits.

6. Settling Strict Contentions:

Theorophy tends to strict contentions by empowering a nuanced and compassionate comprehension of different strict viewpoints. Many contentions emerge from mistaken assumptions, misinterpretations, and dug in creeds. Theorophy's way to deal with compromise includes encouraging a feeling of participation, stressing shared values, and featuring the shared belief that joins devotees. By working with exchange and compromise, Theorophy adds to the goal of strains among strict networks.

7. Instruction for Strict Education:

Training assumes a significant part in advancing interfaith comprehension, and Theorophy advocates for the consideration of the investigation of world religions in instructive educational programs. By cultivating strict proficiency, people gain a more profound appreciation for the convictions and practices of various religions. This information fills in as an establishment for dissipating generalizations, lessening bias, and sustaining a more comprehensive and informed society.

8. Administration and Job Displaying:

Strict pioneers, as powerful figures inside their networks, have an obligation to display resilience, empathy, and participation. Theorophy urges strict pioneers to effectively participate in interfaith drives, cooperative tasks, and joint endeavors that address cultural difficulties. Through initiative and job demonstrating, strict pioneers add to a culture of understanding and concordance inside and past their particular strict networks.

9. Orientation Value and Inclusivity:

Theorophy perceives the significance of orientation value inside conversations of interfaith comprehension. In numerous strict customs, issues of orientation fairness have been wellsprings of pressure. By advancing inclusivity and recognizing the different points of view of people, Theorophy adds to a more adjusted and delegate exchange. This inclusivity cultivates a climate where the voices, everything being equal, paying little mind to orientation, are heard and esteemed.

10. Ecological Stewardship as a Binding together Reason:
    Ecological worries give a shared view to strict networks to team up. The interconnectedness of all life, a topic present in numerous otherworldly practices, lines up with the developing consciousness of natural reliance. The common obligation regarding natural stewardship turns into a binding together power that rises above strict contrasts, provoking devotees to cooperate for the prosperity of the planet. Theorophy empowers the investigation of shared values connected with natural morals, cultivating cooperation on issues of worldwide importance.
11. Accentuation on General Morals:
    Theorophy lines up with the vision of a worldwide ethic, as expressed by Hans Küng. The improvement of a fundamental arrangement of moral rules that can be embraced by people of all strict and philosophical foundations turns into a bringing together power. Theorophy highlights the significance of perceiving all inclusive morals that rise above strict particularities, giving a premise to shared regard and joint effort.
12. Local area Commitment and Compassionate Tasks:
    Interfaith cooperation isn't bound to religious conversations yet reaches out to useful commitment to philanthropic tasks and civil rights drives. Theorophy urges strict networks to meet up in tending to shared cultural difficulties, like neediness, disparity, and medical care. By working one next to the other on projects that benefit the more extensive local area, devotees show the unmistakable effect of interfaith comprehension and participation.
13. Empowering Individual Otherworldly Investigation:

Theorophy puts esteem on individual otherworldly encounters and individual excursions. By empowering people to investigate their honest, own profound ways and encouraging a climate of shared regard for different articulations, Theorophy adds to the making of a more comprehensive and open minded society. This accentuation on private investigation lines up with the possibility that profound development is an exceptional and continuous excursion for every person.

All in all, Theorophy arises as a strong power in advancing interfaith comprehension and concordance by stressing shared otherworldly standards, empowering exchange, embracing strict variety, and tending to clashes through participation. Through its obligation to training, initiative, inclusivity, and commitment to cultural issues, Theorophy fills in as an impetus for building spans among different strict networks. The standards of Theorophy offer a guide for people and social orders to explore the intricacies of strict variety with sympathy, regard, and a common obligation to an amicable concurrence.

Interfaith comprehension and concordance address fundamental yearnings in a world portrayed by different strict practices and convictions. The quest for solidarity in the midst of strict variety isn't just an ethical goal yet in addition a viable need for encouraging tranquil concurrence and tending to worldwide difficulties. This investigation dives into the diverse elements of interfaith comprehension and congruity, inspecting the key standards, difficulties, and approaches that add to the acknowledgment of an additional comprehensive and helpful world.

Underpinnings of Interfaith Comprehension:

At the core of interfaith grasping untruths the acknowledgment of shared human qualities and the affirmation of the inherent pride of each and every person. Interfaith discourse is grounded in the grasping that, in spite of religious contrasts, there are widespread rules that tight spot mankind together. Empathy, equity, love, and the quest for truth are among the ongoing ideas that wind through different strict practices.

Interfaith seeing likewise includes valuing the one of a kind social and verifiable settings that shape every strict custom. It requires an eagerness to take part in deferential discourse, to tune in with compassion, and to perceive the lavishness that variety brings to the human experience. By cultivating a receptive methodology, people can rise above generalizations and biases, taking into consideration a more profound enthusiasm for the complex embroidery of worldwide otherworldliness.

Difficulties to Interfaith Comprehension:

While the objective of interfaith comprehension is honorable, it isn't without challenges. Well established biases, authentic complaints, and misconceptions can make boundaries to significant discourse. The misappropriation of strict lessons for political or philosophical purposes further muddles endeavors to fabricate spans among assorted confidence networks.

Fanaticism and exclusivism inside strict customs can present critical snags to interfaith congruity. A few people might hold unbending convictions that their confidence is the sole vault of truth, prompting a reluctance to take part in certified discourse. Defeating these moves requires a promise to separating generalizations, cultivating common regard, and tending to the main drivers of strict strain.

The Job of Training in Interfaith Comprehension:

Training arises as an amazing asset for advancing interfaith comprehension. By integrating the investigation of world religions into instructive educational plans, social orders can furnish people with the information and social proficiency important to explore a different worldwide scene. Instructive drives ought to underline the philosophical parts of various beliefs as well as their authentic, social, and social settings.

Advancing strict education and intercultural capability since the beginning checks inclinations and generalizations, cultivating a more educated and open minded populace.

Instructive establishments become spaces where understudies learn about their own confidence as well as about the convictions and practices of others. This essential comprehension lays the preparation for people in the future to move toward strict variety with interest and regard.

Interfaith Exchange: A Way to Concordance:
Vital to the advancement of interfaith comprehension is the act of interfaith exchange. Discourse gives an organized and purposeful space for people from various strict foundations to participate in significant discussions. These discoursed can take different structures, including formal conversations, gatherings, and cooperative tasks that address normal difficulties.

Interfaith discourse isn't tied in with looking for religious arrangement yet rather about encouraging shared regard and understanding. Through exchange, members have the chance to share their points of view, explain confusions, and investigate the common qualities that support their convictions. Such collaborations add to the structure of individual connections that rise above strict names, making an establishment for enduring concordance.

The Significance of Strict Forerunners together as one:
Strict pioneers, as compelling figures inside their networks, assume a urgent part in molding mentalities toward interfaith comprehension. Their administration establishes the vibe for how networks approach variety and draw in with individuals from different beliefs. Strict pioneers who model resistance, participation, and a guarantee to discourse establish conditions helpful for interfaith congruity.

Interfaith cooperation among strict pioneers isn't restricted to hypothetical conversations yet reaches out to joint endeavors on helpful activities, civil rights drives, and local area administration. These pragmatic articulations of shared values effectively exhibit the conceivable outcomes of participation and cultivate a feeling of normal reason among different strict networks.

Tending to Strict Contentions Through Interfaith Drives:
Interfaith comprehension turns out to be especially vital in areas where strict contentions have generally caused strain and savagery. Theorophy, an interdisciplinary methodology that looks to uncover general insights across strict practices, can be instrumental in tending to well established hostilities. By stressing shared standards and values, Theorophy adds to compromise endeavors and advances that strict variety can be a wellspring of solidarity as opposed to division.

Interfaith drives focused on compromise include uniting agents from clashing strict gatherings to take part in discourse, intercession, and cooperative critical thinking. The objective is to construct trust, scatter confusions, and figure out some shared interest that rises above verifiable complaints. Such drives require supported exertion, as they explore the intricacies of character, power elements, and international contemplations interlaced with strict contentions.

Orientation Fairness and Interfaith Concordance:
A frequently disregarded part of interfaith comprehension is the job of orientation balance inside strict customs. Numerous religions have wrestled with issues connected with the status and privileges of ladies. Interfaith discourse that remembers conversations for orientation balance adds to a more complete comprehension of the different points of view inside strict networks.

By recognizing and tending to orientation variations, interfaith drives can advance inclusivity and cultivate congruity. Endeavors to guarantee equivalent support of ladies in interfaith discourse, influential positions, and dynamic cycles add to a more delegate and adjusted way to deal with tending to shared difficulties.

Natural Stewardship as a Bringing together Reason:

Natural worries give a novel and intense road for interfaith joint effort. The common acknowledgment of the interconnectedness of all life, present in numerous otherworldly practices, lines up with the developing worldwide attention to environmental relationship. Interfaith drives zeroed in on ecological stewardship rise above strict contrasts, provoking adherents to cooperate for the prosperity of the planet.

Cooperative undertakings tending to environmental change, preservation, and supportable advancement become stages for exhibiting shared values in real life. The ecological development fills in as a shared view that joins people from different strict foundations, underscoring the aggregate liability to safeguard the Earth for people in the future.

The Vision of a Worldwide Ethic:

Hans Küng's vision of a worldwide ethic gives a philosophical establishment to interfaith comprehension and congruity. The thought is to create a fundamental arrangement of moral rules that can be embraced by people of all strict and philosophical foundations. A worldwide ethic underscores values like peacefulness, sympathy, and a pledge to civil rights, giving a common moral system that rises above strict particularities.

The idea of a worldwide ethic provokes people and networks to ponder their common obligations in a quickly globalizing world. By underlining moral rules that are generally pertinent, a worldwide ethic adds to the development of a more amicable and helpful worldwide society.

Interfaith Grasping By and by: Contextual analyses:

A few certifiable models represent the reasonable utilization of interfaith comprehension and concordance. The Parliament of the World's Religions, established in 1893, fills in as a worldwide stage for encouraging interfaith exchange, participation, and cooperative activity. The Parliament unites agents from different strict practices to address worldwide difficulties and advance the upsides of harmony, equity, and manageability.

In the Unified Middle Easterner Emirates, the foundation of the Abrahamic Family House is an outstanding drive advancing interfaith comprehension. This undertaking incorporates a mosque, a congregation, and a gathering place on a similar grounds, representing the common Abrahamic foundations of Islam, Christianity, and Judaism. The Abrahamic Family House fills in as an actual sign of solidarity and common regard among different strict networks.

The Contract for Sympathy, motivated by strict researcher Karen Armstrong's vision, is one more illustration of a worldwide drive encouraging interfaith comprehension. The Sanction calls for people and networks to exemplify the guideline of sympathy in their activities and associations. By empowering a guarantee to sympathy and thoughtfulness, the Contract for Empathy rises above strict limits, advancing a common ethic that reverberates across different social and profound settings.

Chapter 7

Theorophy and Science

The connection among reasoning and science is complicated and multi-layered, traversing hundreds of years of scholarly request and molding the manner in which people see and figure out the world. The two disciplines share a typical heritage, established in mankind's natural interest and want to understand the secrets of presence. Regardless of their common starting points, reasoning and science have advanced particular techniques, objectives, and domains of request, yet they proceed to converge and impact each other in significant ways.

Reasoning, as the antiquated love of shrewdness, originates before the formalization of logical techniques. In the earliest ages of human idea, rationalists wrestled with key inquiries concerning the idea of the real world, cognizance, and presence. These early requests laid the basis for ensuing philosophical customs, molding the scholarly scene for quite a long time into the future. From antiquated Greece toward the Eastern ways of thinking of Confucianism and Taoism, individuals looked to comprehend the fundamental standards overseeing the universe.

As social orders advanced, so did the intricacy of philosophical talk. Greek savants like Thales, Anaximander, and Heraclitus considered the crucial substance of the universe, laying the foundation for mysticism. Socrates, Plato, and Aristotle dug into morals, epistemology, and political way of thinking, establishing the groundwork for efficient request. These masterminds wrestled with questions that rose above the exact perceptions of the regular world, making way for the unmistakable way that way of thinking would take.

The Logical Upset of the seventeenth century denoted an extraordinary period in mankind's set of experiences, as masterminds like Galileo Galilei, Johannes Kepler, and Isaac Newton reformed the comprehension of the actual universe. This time saw the development of experimental techniques and precise perception, signs of the logical strategy. The logical upheaval was not a takeoff from reasoning but rather a uniqueness — a particular way to deal with grasping the world through observational examination.

Notwithstanding this division, the domains of reasoning and science remained interlaced. The Edification time, with illuminators like Immanuel Kant, tried to accommodate reason with experience, encouraging a restored interest in power and epistemology. Kant's supernatural optimism, while tending to the limits of human information, highlighted the significance of philosophical request in outlining the limits and nature of logical information.

The connection among reasoning and science is described by a powerful interchange among hypothesis and experimental examination. Reasoning poses inquiries that might lie past the prompt reach of logical request, examining the idea of the real world, cognizance, and presence. Science, then again, is worried about deliberate perception, trial and error, and the plan of testable speculations to make sense of and foresee regular peculiarities.

Transcendentalism, the part of reasoning worried about the idea of the real world, brings up issues that reach out past the extent of observational science. Inquiries regarding a definitive nature of presence, the brain body issue, and the idea of causation are philosophical in nature, provoking reflection on ideas that may not be straightforwardly perceptible or quantifiable. While science expects to give clarifications in light of exact proof, mystical requests investigate the principal idea of reality itself.

Epistemology, one more key part of reasoning, digs into the nature and cutoff points of human information. It resolves inquiries regarding the idea of truth, avocation, and conviction. Science, as a technique for request, depends on epistemological establishments — suspicions about the dependability of tactile insight, the legitimacy of inductive thinking, and the intelligibility of logical hypotheses. Philosophical reflection on the idea of information impacts the logical strategy and the translation of logical outcomes.

Morals, the part of reasoning worried about moral standards and values, assumes a significant part in molding the moral components of logical request. The turn of events and use of logical information bring up moral issues about the mindful utilization of innovation, the effect on the climate, and the ramifications for human prosperity. Philosophical contemplations illuminate moral structures that guide logical examination and innovative headways.

The convergence of reasoning and science is obvious in the way of thinking of science — a discipline that looks at the suspicions, establishments, and ramifications of logical practice. Scholars of science investigate inquiries concerning the idea of logical clarification, the boundary among science and pseudoscience, and the connection among hypothesis and perception. This intelligent element of reasoning gives a basic focal point through which researchers can assess and refine their techniques.

Hypotheses of logical clarification, for example, Karl Popper's falsifiability measure and Thomas Kuhn's changes in outlook, embody the philosophical examination applied to logical practice. Popper contended that a logical hypothesis should be falsifiable — equipped for being tried and possibly disproven — to be thought of as legitimate. Kuhn, then again, recommended that logical advancement happens through shifts in standards, with times of ordinary science accentuated by progressive changes in understanding.

The way of thinking of science additionally addresses the idea of logical authenticity — the possibility that logical speculations give valid or roughly evident depictions of the world. Banters between logical pragmatists and enemies of pragmatists reflect philosophical investigations into the ontological status of logical elements and the epistemic unwavering quality of logical hypotheses. These discussions feature the complex connection between what science uncovers about the world and how rationalists decipher and conceptualize those disclosures.

Reasoning, as a discipline, adds to the groundworks of logical request by inspecting the presumptions, standards, and ramifications of logical practice. The logical technique itself depends on philosophical presuppositions about the idea of the real world, the unwavering quality of perception, and the legitimacy of inductive thinking. The convergence of reasoning and science isn't just authentic however is a continuous discourse that shapes the direction of the two disciplines.

Thusly, science affects reasoning by giving new observational information and testing conventional philosophical presumptions. Speculations in material science, science, neuroscience, and other logical areas might provoke scholars to reexamine powerful positions, epistemological systems, and moral hypotheses. The revelation of peculiarities, for example, quantum ensnarement or the brain premise of awareness challenges philosophical originations of causation, reality, and the psyche body relationship.

One of the essential issues of crossing point among reasoning and science is the way of thinking of psyche. Inquiries concerning awareness, personality, and the idea of mental states overcome any issues between the experimental investigation of the cerebrum and philosophical request. The brain body issue, which investigates the connection between mental peculiarities and actual cycles, has been a constant and complex test that draws in the two scholars and researchers.

Neuroscience, with its advances in cerebrum imaging and figuring out brain associates of cognizance, contributes observational information that illuminate the philosophical talk on the idea of psyche. Thinkers, thusly, examine the calculated and magical ramifications of neuroscientific discoveries. The reconciliation of logical proof and philosophical investigation is fundamental for fostering an exhaustive comprehension of cognizance, mindfulness, and the emotional parts of human experience.

The way of thinking of language likewise assumes a pivotal part in the convergence of reasoning and science. The clearness and accuracy of language are fundamental for forming logical speculations, conveying exploratory outcomes, and laying out a typical system for logical talk. Philosophical investigations into the idea of language, significance, and correspondence give bits of knowledge into the groundworks of logical language and the difficulties of precisely conveying complex thoughts.

Insightful way of thinking, with its accentuation on sensible examination and semantic accuracy, has especially affected the way of thinking of science. Logicians, for example, Ludwig Wittgenstein and Bertrand Russell investigated the connection among language and reality, preparing for a semantic turn in way of thinking. This phonetic turn had suggestions for how logical speculations are planned, deciphered, and conveyed inside established researchers.

While reasoning and science frequently seek after particular inquiries and systems, they share a typical obligation to objective request and the quest for information. The two disciplines expect to grasp the world, though through various means. Reasoning takes part in applied examination, intelligent reasoning, and theoretical request, while science utilizes exact perception, trial and error, and precise examination.

The polarity among induction and logic, which has saturated the historical backdrop of reasoning, reflects various ways to deal with securing information. Empiricists contend that information is gotten from tactile experience, while realists stress the job of reason and natural ideas. The combination of these points of view is clear in contemporary logical practice, which coordinates experimental proof with judicious conjecturing to develop a far reaching comprehension of the normal world.

The logical technique itself encapsulates a type of applied way of thinking. The most common way of forming speculations, leading investigations, and refining hypotheses in light of observational proof mirrors a guarantee to orderly request and the dismissal of doctrine. The guideline of falsifiability, supported by rationalists like Karl Popper, highlights the significance of exposing logical speculations to thorough testing and expected invalidation.

The connection among reasoning and science reaches out past the domains of hypothetical request and approach to include more extensive cultural ramifications. The moral elements of logical headways bring up issues about the capable utilization of innovation, natural maintainability, and the expected outcomes of logical disclosures. Philosophical morals gives a structure to pondering on these issues, directing researchers and policymakers in pursuing informed and ethically sound choices.

The appearance of contemporary issues, like bioethics, man-made reasoning morals, and natural morals, highlights the continuous significance of philosophical appearance even with quick mechanical and logical advancement. Inquiries concerning the ethical ramifications of hereditary designing, the moral utilization of computerized reasoning, and the obligations of researchers in addressing worldwide moves require philosophical commitment to explore the complex moral scene.

The way of thinking of innovation likewise arises as a basic convergence among reasoning and science. As science and innovation advance, philosophical investigations into the nature and effect of innovation become progressively important.

Crafted by thinkers like Martin Heidegger, Jacques Ellul, and Albert Borgmann investigate the existential, social, and moral components of innovation, bringing up issues about its job in forming human life and the cultural ramifications of mechanical advancement.

Heidegger's idea of "enframing" and Ellul's investigation of the "mechanical society" welcome reflection on the manners by which innovation impacts human discernment, values, and the pith of being. These philosophical bits of knowledge add to a more profound comprehension of the connection between logical development and the human experience, provoking basic assessment of the qualities implanted in mechanical turn of events.

The interdisciplinary idea of contemporary examination further hazy spots the limits among reasoning and science. Fields like mental science, neurophilosophy, and reasoning of science represent the intermingling of philosophical request with experimental examination. Savants team up with researchers to investigate inquiries at the crossing point of the psyche and mind, the idea of cognizance, and the ramifications of developmental hypothesis for how we might interpret life and discernment.

The way of thinking of science likewise draws in with the social components of logical practice. Humanistic and authentic points of view on science inspect the social, political, and financial elements that impact the advancement of logical hypotheses and the acknowledgment of logical ideal models. Crafted by researchers like Thomas Kuhn and Michel Foucault features the social development of logical information and the job of force elements in forming the logical talk.

The impact of values and cultural setting on logical request is a topic that reverberates in both way of thinking and science studies. The affirmation that logical practice isn't esteem nonpartisan difficulties the customary perspective on science as a goal, separated quest for truth. Moral contemplations, social predispositions, and social impacts shape the inquiries researchers pose, the speculations they form, and the manners by which information is deciphered and imparted.

The way of thinking of science likewise faces inquiries concerning the boundary among science and pseudoscience. The models for recognizing genuine logical undertakings from pseudo-logical cases include philosophical contemplations about the idea of proof, the logical strategy, and the dependability of logical hypotheses. This boundary issue mirrors the continuous requirement for philosophical examination in characterizing the limits of authentic logical request.

The connection among reasoning and science is dynamic, described by common impact, discourse, and continuous cooperation. As logical information grows, new philosophical inquiries emerge, provoking existing structures and inciting thinkers to adjust their calculated apparatuses.

On the other hand, philosophical experiences proceed to guide and shape the direction of logical request, cultivating a harmonious relationship that improves the two disciplines.

All in all, the connection among reasoning and science is a rich embroidery woven through the texture of human scholarly history. From the speculative requests of old logicians to the experimental thoroughness of present day science, these disciplines have advanced in equal, each adding to how we might interpret the world in novel ways. While they might veer in techniques and concentration, reasoning and science stay participated in a powerful exchange, impacting and improving each other as they together investigate the secrets of presence.

7.1 Exploring the relationship between Theorophy and scientific inquiry.

The connection among theosophy and logical request is a complex and nuanced interaction that traverses the domains of otherworldliness, power, and experimental perception. Theosophy, established in obscure insight and otherworldly customs, looks to grasp the idea of reality through profound knowledge and extraordinary encounters. Logical request, then again, depends on exact proof, precise perception, and the logical strategy to investigate and make sense of the regular world. In spite of their evident contrasts, the crossing point of theosophy and logical request uncovers a unique relationship that brings up issues about the limits of information, the idea of cognizance, and the manners by which these two methodologies can supplement or challenge one another.

At its center, theosophy is a profound way of thinking that follows its underlying foundations to the late nineteenth 100 years, with figures like Helena Petrovna Blavatsky and the Theosophical Society assuming vital parts in its turn of events. Theosophy includes a wide scope of elusive lessons, drawing from Eastern and Western enchanted customs, old insight, and otherworldly bits of knowledge. Key to theosophical idea is the confidence in an otherworldly, bringing together rule that underlies and associates all parts of presence. This standard is frequently alluded to as the "Heavenly," "Outright," or "One."

Theosophy states that past the noticeable and quantifiable parts of reality lies a more profound, otherworldly aspect that can be caught through instinctive and magical encounters. The investigation of this secret reality includes digging into the idea of awareness, the reason for presence, and the interconnectedness of all life. Theosophical lessons place the presence of otherworldly progressive systems, grandiose regulations, and an all inclusive transformative cycle that directs the improvement of cognizance through progressive stages.

Logical request, then again, works inside a structure of experimental perception and testable theories.

The logical strategy, with its accentuation on efficient perception, trial and error, and the detailing of falsifiable speculations, has been strikingly fruitful in unwinding the secrets of the normal world. Science, as a discipline, looks to give clarifications to discernible peculiarities in view of proof that can be freely confirmed and recreated.

The pressure among theosophy and logical request emerges from the varying systems and epistemological underpinnings of these two methodologies. While theosophy depends on otherworldly knowledge, mysterious encounters, and exclusive insight, science requests experimental proof and adherence to the standards of systemic naturalism. The test lies in accommodating these different strategies and deciding the degree to which they can coincide or illuminate one another.

One mark of convergence among theosophy and logical request is the investigation of cognizance. Theosophical lessons set that cognizance isn't bound to the actual body however reaches out past it, interfacing people to a more extensive, infinite awareness. This point of view difficulties the regular logical comprehension of awareness as an emanant property of the cerebrum's brain action. Theosophy proposes that cognizance is key and goes before the material sign of the actual body.

Lately, the logical investigation of cognizance has extended, integrating interdisciplinary methodologies that draw from neuroscience, brain science, and reasoning. While researchers dominatingly center around the brain connects of cognizance and the mind's job in creating emotional encounters, theosophy empowers a more extensive investigation of awareness that envelops otherworldly aspects and extraordinary conditions of mindfulness. The exchange between these viewpoints opens roads for a more exhaustive comprehension of the mind boggling nature of cognizance.

One more area of convergence is the investigation of the idea of the real world and the interconnectedness of all presence. Theosophy sets a comprehensive perspective in which everything is interconnected and part of a general, grandiose request. This viewpoint reverberations subjects tracked down in Eastern ways of thinking and supernatural customs, underscoring the solidarity of all life and the reliance of each and every person with the more noteworthy entirety.

Conversely, logical request has customarily taken on a reductionist methodology, separating complex peculiarities into more modest, more reasonable parts for investigation. In any case, progressions in fields like frameworks hypothesis, environment, and quantum material science have provoked a shift towards perceiving the interconnectedness of frameworks at different levels. The arising field of natural science, for instance, recognizes the multifaceted snare of connections among organic entities and their surroundings, lining up with the theosophical accentuation on interconnectedness.

The investigation of awareness and the interconnectedness of reality likewise prompts inquiries regarding the idea of insight and the restrictions of human detects. Theosophy recommends that human discernment is restricted to the actual faculties and that genuine comprehension requires rising above these impediments through otherworldly knowledge and internal arousing. Interestingly, science depends on the observational information given by the faculties and innovative instruments to concentrate on the outside world.

Progressions in fields like neuroscience and brain research, be that as it may, have enlightened the intricacies of discernment and perception. The investigation of perceptual deceptions, mental predispositions, and changed conditions of cognizance challenges the idea of a direct, objective reality. This combination of points of view welcomes a nuanced exchange about the idea of discernment, the job of cognizance in molding reality, and the potential for growing our grasping past ordinary tangible encounters.

Theosophy's accentuation on otherworldly advancement and the improvement of higher resources of cognizance additionally meets with contemporary conversations in mainstream researchers. While theosophy discusses the development of awareness through progressive stages, science investigates the advancement of life structures through natural and hereditary cycles. The idea of "transhumanism," which imagines the upgrade of human abilities through mechanical means, reverberates with the theosophical thought of otherworldly advancement and the development of inert human possibilities.

The connection among theosophy and logical request turns out to be especially captivating while thinking about the investigation of unpretentious energies, clairvoyant peculiarities, and the brain body association. Theosophy sets the presence of unpretentious energies and profound powers that impact the physical and mental parts of people. Practices, for example, reflection, energy mending, and the enlivening of clairvoyant resources are basic to the theosophical way, meaning to adjust people to these inconspicuous elements of presence.

According to a logical point of view, the investigation of unobtrusive energies and mystic peculiarities has frequently been met with suspicion because of the difficulties of exact check and the potential for pseudo-logical cases. In any case, the developing interest as a main priority body connections, psychoneuroimmunology, and a self-influenced consequence highlights the interconnectedness among mental and actual prosperity. The investigation of correlative and elective medication likewise mirrors an eagerness inside logical circles to examine peculiarities that may not fit customary standards.

Equals can be drawn between the theosophical idea of the air — an energy field encompassing living creatures — and logical examination on bioelectromagnetism. While theosophy depicts the emanation as an indication of profound energies and conditions of cognizance, bioelectromagnetism investigates the electrical and attractive fields created by living life forms.

The discourse between these points of view brings up issues about the idea of energy, its effect on wellbeing and cognizance, and the potential for incorporating profound bits of knowledge with logical comprehension.

Theosophy's commitment with cosmology and the idea of the universe additionally meets with logical investigations into the beginnings and design of the universe. Theosophical lessons propose a repetitive model of grandiose development, with exchanging times of sign and disintegration. This repetitive viewpoint appears differently in relation to the direct cosmological models got from logical perceptions of the extending universe and the hypothesis of the Enormous detonation.

The discourse among theosophy and cosmology prompts reflection on the idea of time, the starting points of the universe, and the recurrent examples saw in vast peculiarities. Logical revelations, for example, grandiose expansion and the disclosure of dull energy, challenge existing cosmological standards and welcome a reexamination of the essential standards overseeing the universe. The convergence of theosophical cosmology with state of the art astronomy brings up captivating issues about the idea of the real world, the motivation behind presence, and a definitive predetermination of the universe.

While the connection among theosophy and logical request offers rich ground for investigation, recognizing the difficulties and strains that exist between these two approaches is fundamental. Theosophy's dependence on otherworldly understanding, mysterious encounters, and obscure lessons can be seen as contradictory with the thorough guidelines of observational proof requested by the logical technique. The absence of replicability and the emotional idea of enchanted encounters present troubles for incorporation into the logical structure.

7.2 Discussing how Theorophical principles align with certain scientific concepts.

The arrangement between theosophical standards and certain logical ideas offers a captivating investigation into the convergences of otherworldliness, mysticism, and exact request. Theosophy, with its foundations in obscure insight and supernatural customs, places a far reaching perspective that envelops profound development, interconnectedness, and the investigation of stowed away elements of the real world. While theosophical lessons frequently utilize emblematic language and draw from assorted social and strict customs, there are charming equals between these elusive standards and explicit logical ideas. This conversation dives into a few key regions where theosophical standards line up with logical thoughts, offering bits of knowledge into the possible combination of otherworldly insight and experimental comprehension.

One major area of arrangement lies in the theosophical idea of a widespread, binding together rule — the Heavenly, Outright, or One — which underlies and interfaces all parts of presence.

This thought tracks down reverberation in logical conversations about a crucial solidarity in the universe. In material science, the quest for a brought together hypothesis that can portray the basic powers of the universe, like gravity and electromagnetism, mirrors a mission for a hidden, bringing together standard. The quest for a "hypothesis of everything" in physical science repeats the theosophical idea of an enormous solidarity that rises above the evident variety of the material world.

The theosophical accentuation on otherworldly development lines up with specific parts of transformative science. Theosophy sets a persistent, deliberate unfoldment of cognizance through progressive stages, reflecting the transformative cycle depicted in science. While theosophical advancement stretches out past the actual domain to incorporate profound turn of events, the possibility of a continuous and intentional development imparts shared view to the logical comprehension of the steady improvement of life structures throughout land time scales.

Besides, theosophical lessons on the interconnectedness of all life line up with natural and frameworks thinking in science. Theosophy recommends that each individual is essential for a more noteworthy entire, and activities have repercussions that resonate all through the interconnected snare of presence. This interconnected perspective reverberates with biological science, which perceives the multifaceted connections among life forms and their surroundings. Ideas like environments, input circles, and the fragile equilibrium of nature mirror a logical appreciation for the interconnectedness that theosophy likewise underscores.

The investigation of cognizance gives one more mark of arrangement between theosophical standards and logical request. Theosophy attests that awareness isn't restricted to the actual body and that genuine comprehension requires rising above tangible impediments. This point of view lines up with the developing field of cognizance concentrates inside neuroscience and brain research, which tries to unwind the secrets of abstract insight and the idea of mindfulness. While theosophy approaches cognizance from an otherworldly and supernatural point, the logical examination of brain relates of cognizance and modified conditions of mindfulness combines with theosophical experiences.

Theosophy's idea of unobtrusive energies, for example, those related with the quality, lines up with specific logical examinations concerning bioelectromagnetism. While theosophy depicts the air as a sign of otherworldly energies and conditions of cognizance, bioelectromagnetism investigates the electrical and attractive fields created by living creatures. This combination brings up issues about the idea of energy, its effect on wellbeing and awareness, and the potential for incorporating otherworldly experiences with logical comprehension.

The theosophical viewpoint on cosmology, including repeating models of vast advancement, finds reverberation in the developing field of astronomy. Logical revelations testing conventional cosmological models, like enormous expansion and dull energy, open up exchanges about the idea of time, the beginnings of the universe, and the recurrent examples saw in grandiose

peculiarities. While theosophical cosmology is communicated in emblematic and metaphorical terms, the arrangement with specific parts of contemporary cosmology welcomes thought on the more extensive nature of the real world.

Additionally, the theosophical accentuation on all encompassing prosperity and the brain body association lines up with developing patterns in integrative medication and psychoncuroimmunology. Theosophy recommends that actual wellbeing is unpredictably associated with profound and mental prosperity, mirroring a methodology that thinks about the entire individual. Logical examinations investigating the exchange between mental elements, the sensory system, and the safe framework resound with the comprehensive point of view tracked down in theosophical lessons.

The investigation of the psyche body association likewise meets with the logical examination of care practices and contemplation. Theosophy energizes the development of higher conditions of awareness through insightful works on, lining up with the developing collection of logical exploration on the advantages of care for emotional well-being, stress decrease, and generally prosperity. The combination of theosophical bits of knowledge with logical discoveries in this domain features the potential for a correlative way to deal with figuring out the significant effect of cognizance on the body.

Notwithstanding these areas of arrangement, perceiving the differentiations and likely difficulties in accommodating theosophical standards with specific logical concepts is critical. Theosophy frequently utilizes emblematic language, figurative accounts, and otherworldly structures that vary from the exact and quantitative language of logical talk. The emotional idea of otherworldly encounters and the absence of experimental evidence in theosophical lessons can present difficulties for mix into the observational and objective guidelines of logical request.

Furthermore, the accentuation on profound understanding, exclusive insight, and the quest for extraordinary information in theosophy may not promptly line up with the systemic naturalism inborn in logical examinations. While theosophy embraces stowed away elements of reality available through profound instinct, science regularly centers around recognizable, quantifiable, and replicable peculiarities.

The connection among theosophy and science, in this manner, might be described more by discourse and shared investigation than by a consistent coordination of techniques. The continuous discussion between these viewpoints welcomes researchers, specialists, and scholars from the two domains to take part in a deferential trade of thoughts, perceiving the potential for enhancement and development of understanding that can emerge from their combination.

All in all, the arrangement between theosophical standards and certain logical ideas gives a rich territory to investigation at the crossing point of otherworldliness and observational request. Theosophy, with its obscure insight and enchanted customs, shares captivating equals with logical thoughts in regions like the solidarity of presence, otherworldly development, cognizance, interconnectedness, and cosmology. While theosophical lessons frequently utilize representative and figurative language, and stress otherworldly bits of knowledge, the possible combination with logical ideas recommends a powerful transaction that can add to a more thorough comprehension of the idea of the real world. The continuous exchange among theosophy and science holds the commitment of encouraging a comprehensive way to deal with information — one that coordinates both profound insight and exact investigation in the continuous mission to unwind the secrets of presence.

7.3 Addressing the potential for a harmonious integration of spirituality and science.

The investigation of the potential for an agreeable joining of otherworldliness and science is a multi-layered venture that includes accommodating different perspectives, procedures, and ways to deal with grasping the idea of the real world. Otherworldliness, frequently established in strict customs, supernatural experiences, and scrutinizing rehearses, envelops the mission for significance, reason, and a more profound association with the extraordinary. Science, then again, depends on exact proof, precise perception, and the logical technique to unwind the secrets of the regular world. While customarily seen as discrete spaces, the chance of coordinating otherworldliness and science has acquired expanding consideration lately. This conversation digs into different parts of this combination, investigating areas of union, challenges, and the potential for a blend that enhances the two viewpoints.

One road for coordination lies in perceiving the correlative idea of otherworldliness and science in tending to various components of human experience. Otherworldliness frequently dives into inquiries of significance, reason, and the idea of cognizance, offering experiences into the emotional and existential parts of life. Science, with its experimental strategies and precise request, succeeds in investigating the outer, noticeable parts of the material world. By recognizing the particular yet reciprocal jobs of these spaces, an amicable mix becomes conceivable — one that embraces both the internal elements of human experience and the outside investigation of the normal world.

In addition, a coordinated methodology can draw motivation from the rich woven artwork of shrewdness tracked down in different otherworldly practices. Numerous otherworldly lessons stress the interconnectedness of all life, the significance of sympathy and compassion, and the quest for internal change to improve society. Coordinating these moral and moral standards into logical undertakings can add to a more comprehensive and socially capable way to deal with research and mechanical progressions. Perceiving the moral ramifications of logical revelations, the effect on the climate, and the prosperity of people lines up with the profound accentuation on moral lead and the interconnectedness, everything being equal.

The investigation of cognizance gives a prolific ground to the combination of otherworldliness and science. Otherworldliness, drawing from scrutinizing customs and enchanted encounters, offers bits of knowledge into adjusted conditions of cognizance, extended mindfulness, and the idea of oneself. While logical ways to deal with cognizance frequently center around the brain relates and noticeable parts of mental peculiarities, there is a developing acknowledgment of the need to integrate emotional encounters and subjective aspects into the investigation of awareness. Incorporating the scrutinizing experiences of profound customs with the exact strategies for neuroscience opens roads for a more extensive comprehension of the many-sided nature of cognizance.

Care rehearses, established in otherworldly practices like Buddhism, represent an expected place of joining among otherworldliness and science. The boundless reception of care in clinical settings and logical examination mirrors a developing acknowledgment of the advantages of scrutinizing rehearses for emotional wellness and prosperity. The combination of care into helpful mediations and mental exploration highlights the potential for a cooperative connection among otherworldliness and science — one that encourages human thriving by integrating old scrutinizing shrewdness into current observational methodologies.

The investigation of inconspicuous energies and the brain body association gives one more pathway to joining. Otherworldly customs frequently talk about unpretentious energies, chakras, and life powers that impact the physical and mental parts of individuals. While these ideas may not fit inside the ordinary logical structure, there are areas of cross-over, like the investigation of bioelectromagnetism and the effect of mental states on physiological cycles. Overcoming any issues between otherworldly bits of knowledge and logical request in this space requires a liberal investigation that looks for shared conviction and recognizes the constraints and capability of the two viewpoints.

The possible joining of otherworldliness and science likewise reaches out to the domain of cosmology. Numerous profound customs offer cosmological accounts that portray the beginnings of the universe, the reason for presence, and the interconnectedness of all creation. While these stories may not line up with logical cosmological models as far as observational unquestionable status, they can move a feeling of marvel, wonderment, and love for the universe. Coordinating the lovely and emblematic parts of cosmological stories with the experimental discoveries of astronomy welcomes a comprehensive commitment with the secrets of the universe.

One more road for combination includes perceiving the restrictions of reductionist and materialistic ideal models in science. Reductionism, the methodology of separating complex peculiarities into easier parts for investigation, has been a foundation of logical request. Nonetheless, this reductionist viewpoint has its limits, particularly while tending to complex frameworks, rising properties, and the subjective parts of human experience.

The coordination of otherworldly experiences into logical talk energizes a more comprehensive and frameworks situated approach — one that perceives the reliance of parts, the rise of novel properties, and the significance of subjective aspects in figuring out complex peculiarities.

Transdisciplinary approaches, which draw from numerous disciplines including otherworldliness, reasoning, and technical studies, offer a system for reconciliation. Transdisciplinary research looks to rise above disciplinary limits, perceiving that perplexing issues frequently require assorted viewpoints and systems. By cultivating coordinated effort and discourse between profound professionals, researchers, logicians, and different partners, transdisciplinary approaches can work with a more nuanced and comprehensive comprehension of intricate peculiarities. The coordination of different viewpoints, including otherworldly bits of knowledge, adds to a more far reaching investigation of the mind boggling embroidery of human experience and the normal world.

While the potential for an agreeable mix of otherworldliness and science is promising, it is vital for address the difficulties that might emerge. One test lies in exploring the variety of otherworldly points of view and keeping away from a reductionist methodology that sums up all profound practices into a particular system. The lavishness of otherworldly variety requires a nuanced and conscious commitment that perceives the special commitments of various practices while looking for ongoing ideas that can add to a mutual perspective.

The potential for opinion, whether in profound or established researchers, represents another test. The receptiveness to joining requires a readiness to rise above inflexible convictions and standards, recognizing the limits and developing nature of both profound and logical understandings. Embracing the soul of request and the acknowledgment that information is dynamic can encourage a more adaptable and versatile way to deal with reconciliation.

Moreover, moral contemplations are fundamental in the joining of otherworldliness and science. The likely abuse of profound experiences for manipulative purposes or the burden of explicit otherworldly convictions in logical settings raises moral worries. It is pivotal to maintain standards of straightforwardness, informed assent, and regard for different viewpoints in any cooperative endeavors among profound and mainstream researchers. Moral rules that advance respectability, inclusivity, and the prosperity of people can assist with exploring the intricate landscape of reconciliation.

The potential for predisposition, whether implied or express, likewise presents difficulties. Both profound and established researchers might bring assumptions, social inclinations, or individual convictions to the reconciliation cycle. Cautiousness in perceiving and tending to predispositions, encouraging variety in viewpoints, and keeping a receptiveness to useful analysis add to a more strong and comprehensive reconciliation of otherworldliness and science.

The advancing idea of language and phrasing presents one more test in the reconciliation cycle. Profound customs frequently utilize representative, metaphorical, or socially unambiguous language that may not promptly line up with the exact and normalized wording of logical talk. Overcoming this issue requires a cautious thought of language, encouraging a common jargon that regards the subtleties of both profound and logical points of view without weakening the wealth of by the same token.

Regardless of these difficulties, the likely advantages of an amicable combination of otherworldliness and science are immense. Such coordination can possibly add to more comprehensive training that praises both the experimental and scrutinizing aspects of human request. Instructive systems that coordinate logical standards with moral contemplations, care rehearses, and philosophical reflections can support balanced people who value the interconnectedness of information and the significance of moral direct in their interests.

The convergence of otherworldliness and science addresses a dynamic and developing discourse that draws in with key inquiries regarding the idea of the real world, human life, and the quest for information. Otherworldliness, frequently established in strict customs, supernatural experiences, and pondering practices, looks to investigate the extraordinary elements of presence, offering a structure for grasping significance, reason, and the interconnectedness of all life. Science, then again, works inside a worldview of exact proof, deliberate perception, and the logical technique, meaning to disentangle the secrets of the regular world through genuine request. Notwithstanding their apparently unique methodologies, the connection among otherworldliness and science is described by a mind boggling exchange of intermingling, disparity, and continuous discourse.

One vital area of investigation lies in the likely combination between otherworldly bits of knowledge and logical comprehension with respect to the idea of awareness. Otherworldliness frequently sets that cognizance isn't bound to the actual body, proposing the presence of an extraordinary and timeless part of oneself. This viewpoint lines up with the enduring inquiry of whether awareness emerges exclusively from brain action, as set by logical realism, or on the other hand assuming that it has aspects that stretch out past the material domain.

Contemporary neuroscience wrestles with the intricacies of cognizance, looking to comprehend the brain relates and components that lead to emotional encounters. While science basically centers around the discernible and quantifiable parts of cognizance, a few researchers and thinkers advocate for a more comprehensive methodology that recognizes the subjective and emotional components of inward experience. This mix of pensive bits of knowledge and emotional aspects into logical talk encourages a more complete investigation of cognizance, including both the goal and abstract features.

Care rehearses, established in profound customs like Buddhism, embody a mark of union among otherworldliness and science. The far reaching reception of care in clinical settings and logical examination mirrors an acknowledgment of the advantages of pondering practices for emotional wellness and prosperity. Logical investigations investigating the effect of care on brain versatility, stress decrease, and mental working show a cooperative energy between otherworldly insight and experimental request. The combination of care into remedial mediations highlights the potential for an amicable joint effort among otherworldliness and science in advancing human thriving.

The investigation of awareness additionally digs into inquiries regarding the idea of self and personality. Otherworldly customs frequently place the presence of a higher or extraordinary self past the egoic character, recommending a more profound and interconnected part of uniqueness. Interestingly, neuroscientific points of view will generally connect the self with the cerebrum's cycles and brain organizations.

However, the combination emerges when one thinks about the ramifications of brain adaptability — the cerebrum's ability to revamp and adjust because of involvement. Logical discoveries recommend that the healthy identity isn't unbendingly attached to explicit brain structures and can be impacted by deliberate mental works on, repeating otherworldly bits of knowledge about the pliability and breadth of awareness.

Additionally, the reconciliation of scrutinizing rehearses, like contemplation, into logical exploration has enlightened the pliancy of the mind and the potential for groundbreaking changes in mindfulness. Concentrates on the impacts of reflection on the mind's design and capability feature the limit with respect to deliberate practices to shape the brain premise of cognizance, offering a gathering point between otherworldly investigation and logical request.

One more area of union lies in the investigation of unpretentious energies and the brain body association. Profound customs frequently discuss unobtrusive energies, life powers, and chakras that impact the physical and mental parts of individuals. While these ideas may not adjust straightforwardly with logical standards, there are areas of cross-over, like the investigation of bioelectromagnetism.

Bioelectromagnetism explores the electrical and attractive fields created by living creatures, recognizing the presence of electromagnetic peculiarities inside the human body. While the logical investigation of bioelectromagnetism may not straightforwardly correspond with profound ideas of inconspicuous energies, the acknowledgment of electromagnetic fields as basic to natural working opens roads for discourse and investigation. This combination prompts thought of the expected interconnectedness between the inconspicuous components of otherworldly lessons and the electromagnetic viewpoints investigated by logical request.

Notwithstanding, challenges emerge in exploring this combination, especially concerning the potential for reductionism or distortion. Reductionism, the propensity to separate complex peculiarities into less difficult parts for examination, may prompt a distorted comprehension of otherworldly ideas when converted into logical terms. Otherworldly customs frequently utilize representative language and allegorical stories that might oppose interpretation into the exact and quantitative language of logical talk. Cautious thought of the subtleties of both otherworldly and logical viewpoints is fundamental to stay away from reductionist methodologies that might subvert the lavishness of every area.

The investigation of the brain body association likewise brings up issues about the expected effect of mental states on physiological cycles. While otherworldliness underlines the interconnectedness of mental, close to home, and actual prosperity, logical investigation into psychoneuroimmunology and a self-influenced consequence perceives the complicated connection between mental states and real wellbeing.

A self-influenced consequence, specifically, features the impact of conviction, assumption, and cognizance on the body's reaction to treatment. Logical examinations on a self-influenced consequence highlight the significance of mental elements in mending, recommending that cognizance assumes a critical part in impacting wellbeing results. This acknowledgment lines up with otherworldly viewpoints that underscore the interconnected idea of psyche and body, opening roads for cooperative investigation that incorporates both profound bits of knowledge and logical discoveries.

The potential for intermingling reaches out to the domain of morals and profound quality. Profound practices frequently give moral systems grounded in standards of sympathy, compassion, and interconnectedness. While science generally takes on a spellbinding as opposed to prescriptive position, there is a developing acknowledgment of the moral ramifications of logical progressions and mechanical developments.

The field of bioethics, for instance, draws in with inquiries regarding the capable utilization of innovation, the moral ramifications of hereditary designing, and the effect of logical revelations on cultural qualities. Otherworldly experiences into the moral components of human activities, natural stewardship, and the interconnectedness of all life can contribute significant points of view to the continuous talk in bioethics and the dependable use of logical information.

Besides, the combination of otherworldliness and science in the moral space prompts contemplations about the cultural outcomes of logical headways. Computerized reasoning, biotechnology, and other arising advancements bring up significant moral issues about protection, independence, and the potential for unseen side-effects. Incorporating profound upsides of sympathy, equity, and veneration for life into moral structures for innovative improvement encourages a more comprehensive and mindful way to deal with logical advancement.

While the potential for union among otherworldliness and science is obvious, recognizing the difficulties and strains that persist is significant. The language obstruction, varying epistemologies, and social variety among profound practices present deterrents to consistent combination. The utilization of figurative language, emblematic symbolism, and allegorical stories in profound lessons might oppose interpretation into the exact and normalized language of logical talk. Overcoming this issue requires a promise to conscious discourse, perceiving the interesting commitments of the two viewpoints without undermining their respectability.

One more test lies in the potential for opinion or protection from novel thoughts inside both profound and established researchers. The receptiveness to mix requires an eagerness to rise above unbending convictions, ideal models, and disciplinary limits. Embracing the soul of request and perceiving that information is dynamic takes into consideration a more adaptable and versatile way to deal with the combination of otherworldliness and science.

Moral contemplations likewise assume a urgent part in exploring the union among otherworldliness and science. The expected abuse of otherworldly bits of knowledge for manipulative purposes, the burden of explicit profound convictions in logical settings, or the disregard of moral contemplations in mechanical progressions raise moral worries. Maintaining standards of straightforwardness, informed assent, and regard for different viewpoints is fundamental in any cooperative endeavors among profound and established researchers.

In spite of these difficulties, the continuous investigation of the assembly among otherworldliness and science holds colossal commitment for advancing the two spaces. An incorporated methodology that respects the extraordinary commitments of every viewpoint can prompt a more far reaching comprehension of the intricacies of human experience and the regular world.

Chapter 8

Mystical Experiences in Theorophy

Mysterious encounters have been a common subject all through mankind's set of experiences, rising above social and strict limits. These significant experiences with the heavenly or a definitive reality have frequently been portrayed as inexpressible, resisting simple clarification or enunciation. Theorophy, a term begat to embody the association of hypothetical and mysterious components, looks to investigate and figure out the supernatural elements of presence through a philosophical focal point.

At its center, Theorophy dives into the idea of the real world, cognizance, and the interaction between the material and irrelevant domains. A scholarly pursuit looks to unwind the secrets of presence, perceiving the limits of simply sane methodologies and embracing the instinctive, experiential parts of human awareness. In the investigation of enchanted encounters inside Theorophy, one experiences a rich embroidery of considerations, convictions, and practices that range hundreds of years and mainlands.

The foundations of Theorophy can be followed back to old philosophical customs that incorporated supernatural quality into their perspective. In the East, the magical encounters of Hinduism, Buddhism, and Taoism have been primary to the advancement of Theorophical thought. The Upanishads, for instance, give bits of knowledge into the idea of oneself (Atman) and its association with a definitive reality (Brahman). Additionally, the otherworldly encounters of Buddhist contemplation and the Taoist comprehension of the Tao as an unutterable wellspring of all presence add to the Theorophical investigation of awareness and the extraordinary.

In the West, the supernatural encounters inside Christian magic, Kabbalah, and Sufism play had a vital impact in molding Theorophy. The works of spiritualists like Meister Eckhart, John of the Cross, and Julian of Norwich offer looks into the extraordinary experiences with the heavenly. The Kabbalistic lessons on the idea of God and the Sufi spiritualists' journey for association with the Darling add to the rich embroidered artwork of Theorophical thought.

The incorporation of otherworldly encounters into Theorophy doesn't look to nullify reason or excuse the significance of scholarly request. All things considered, it perceives the restrictions of a simply rationalistic methodology and recognizes the presence of aspects of reality that rise above the grip of the logical psyche. Theorophy, as a blend of hypothesis and supernatural quality, welcomes people to investigate the significant and frequently unspeakable parts of presence while keeping a pledge to thorough scholarly request.

One of the focal subjects inside Theorophy is the idea of awareness and its relationship to a definitive reality. Mysterious encounters frequently include an increased condition of mindfulness or modified cognizance, driving people to see real factors past the customary tactile encounters. Theorophy investigates the possibility that cognizance isn't restricted to the material cerebrum however is an all inclusive and interconnected part of the real world.

In the mysterious customs of different societies, professionals portray encounters of self image disintegration, where the limits among self and other, subject and article, obscure or break up altogether. This disintegration of the inner self is much of the time seen as a passage to a more profound, more significant comprehension of the real world. Theorophy examines whether cognizance, in its most flawless structure, is an otherworldly, bringing together power that underlies all of presence.

In addition, Theorophy wrestles with whether or not otherworldly encounters give veritable experiences into the idea of the real world or on the other hand on the off chance that they are just the result of mental and brain processes. While neuroscience might offer clarifications for specific parts of mysterious encounters, Theorophy recognizes that the emotional, individual nature of these experiences can't be completely caught by logical examination alone. Theorophy welcomes a comprehensive methodology that coordinates both observational examination and thoughtful investigation.

The connection among time and enchanted encounters is one more charming aspect of Theorophy. Numerous spiritualists portray an immortal quality to their experiences with the heavenly, where customary thoughts of past, present, and future become insignificant. Theorophy inspects the ramifications of this immortal aspect, addressing whether it proposes a more profound reality past the fleeting requirements of our regular experience.

Theorophical conversations on time and mystery frequently cross with cosmological requests, pondering the starting points and extreme destiny of the universe. Some Theorophers suggest that magical encounters give looks into an immortal, timeless reality that rises above the constraints of the actual universe. Others investigate the possibility that time itself is a develop of human insight and that enchanted encounters offer an immediate experience with an immortal, ever-present element of the real world.

Theorophy, while established in antiquated customs, likewise draws in with contemporary philosophical and logical points of view. In the exchange among enchantment and quantum material science, for instance, Theorophers look for shared belief between the otherworldly comprehension of interconnectedness and the quantum idea of non-region. The investigation of cognizance in the area of neuroscience converges with Theorophical investigations into the idea of oneself and its association with a definitive reality.

While Theorophy recognizes the worth of scholarly thoroughness and exact examination, it additionally perceives the significance of individual experience and direct commitment with the enchanted components of presence. Theorophers contend that otherworldly encounters, a long way from being simple emotional peculiarities, can act as a wellspring of significant insight and extraordinary understanding. The reconciliation of hypothesis and enchantment, as per Theorophy, offers a more extensive comprehension of the idea of the real world.

Theorophical points of view on supernatural encounters reach out past individual experiences with the heavenly to envelop aggregate and social aspects. The job of otherworldliness in the development of strict and profound practices is a point of convergence of Theorophical request. Theorophy investigates how otherworldly experiences have molded the moral, moral, and social structures of different social orders since the beginning of time.

With regards to coordinated religions, Theorophers question how regulated structures have either worked with or obstructed the outflow of magical encounters. The strain between spiritualists who guarantee immediate, unmediated admittance to the heavenly and strict specialists who look to control and decipher such encounters is a common subject inside Theorophy. Theorophers look at the manners by which magical encounters challenge or support laid out strict teachings, prompting the advancement of strict idea and practice.

Theorophical conversations additionally address the groundbreaking capability of mysterious encounters at the cultural level. Some Theorophers contend that the bits of knowledge acquired through magical experiences have the ability to rouse sympathy, compassion, and a feeling of interconnectedness, encouraging a more agreeable and moral worldwide local area. Others alert against the potential for abuse or distortion of enchanted encounters, stressing the requirement for acumen and moral contemplations in the mix of otherworldly insight into cultural designs.

Moral contemplations inside Theorophy stretch out past the cultural level to the singular's moral and profound turn of events. Magical encounters are frequently portrayed as impetuses for significant individual change, driving people to reexamine their qualities, needs, and feeling of direction. Theorophy investigates the moral ramifications of such groundbreaking encounters, taking into account how they might impact one's relationship with oneself, others, and the more extensive natural setting.

Moreover, Theorophy mulls over the job of magical encounters in the development of ideals like modesty, sympathy, and astuteness. Spiritualists across customs frequently discuss a profound feeling of interconnectedness with all of creation, cultivating an awareness of certain expectations and care for the prosperity of others. Theorophers investigate whether mysterious bits of knowledge can act as an establishment for moral rules that rise above social and strict limits, adding to a more comprehensive and sympathetic worldwide ethic.

Theorophy likewise draws in with the difficulties and discussions encompassing the translation and approval of otherworldly encounters. Cynics might excuse such encounters as simple pipedreams or dreams, crediting them to mental or neurological elements. Theorophers answer by featuring the culturally diverse consistency of enchanted accounts and the groundbreaking effect these encounters can have on people and social orders.

In resolving the topic of approval, Theorophy perceives the intrinsic abstract nature of mysterious encounters. While recognizing the significance of individual genuineness and truthfulness in relating these experiences, Theorophers additionally investigate the job of common and social approval. Theorophical points of view on the public part of magic consider how shared customs, images, and stories add to the approval and translation of supernatural encounters inside a specific social or strict setting.

The convergence of orientation and magic is one more critical component of Theorophy. Theorophers basically inspect verifiable and social portrayals of spiritualists, taking into account how orientation elements impact the translation and acknowledgment of magical encounters. Theorophy looks to enhance voices that have been underestimated or barred from customary accounts, investigating the commitments of ladies spiritualists and the effect of orientation jobs on the articulation and gathering of otherworldly bits of knowledge.

Theorophical conversations on orientation and mystery stretch out past the parallel system, recognizing the variety of orientation characters and articulations. The interconnection of orientation with different parts of character, like race, class, and sexuality, is a focal concentration inside Theorophy. Theorophers investigate how different social and social elements converge to shape the encounters and translations of enchantment among people with assorted characters.

In the journey to coordinate mysterious encounters into Theorophy, the job of language and imagery turns into a focal subject. Mysterious experiences are in many cases portrayed as unutterable, past the limit of customary language to catch or convey the experience completely. Theorophers wrestle with the restrictions of language, perceiving that the actual idea of otherworldly encounters difficulties regular methods of articulation.

Imagery and representation are fundamental apparatuses inside Theorophy to explore the unutterable idea of otherworldly experiences. Theorophers draw upon strict, social, and prototype images to verbalize parts of the enchanted experience that oppose clear portrayal. The utilization of purposeful anecdote and fantasy inside Theorophy conveys the groundbreaking power and significant bits of knowledge that rise up out of magical commitment with a definitive reality.

The connection among mystery and imaginative articulation is additionally investigated inside Theorophy. Craftsmen, writers, and performers frequently draw motivation from enchanted encounters, utilizing their particular mediums to convey the otherworldly and unutterable components of the real world. Theorophers consider how craftsmanship fills in as a scaffold between the supernatural and the ordinary, welcoming people to mull over and draw in with the secrets of presence through tasteful encounters.

With regards to Theorophy, the job of imagery stretches out past creative articulation to envelop the understanding of strict and legendary texts. Theorophers participate in hermeneutical requests, investigating how sacrosanct sacred writings and fantasies encode supernatural experiences and act as guides for those looking to explore the otherworldly way. Theorophy energizes a dynamic and nuanced way to deal with strict texts, perceiving the numerous layers of significance and the potential for extraordinary insight inside these hallowed stories.

Theorophy additionally wrestles with the subject of strict pluralism and the conjunction of assorted magical customs. Theorophers investigate the shared traits and contrasts among different supernatural ways, looking to observe widespread rules that underlie the assorted articulations of the enchanted insight. Theorophy advances a feeling of inclusivity, perceiving the legitimacy of various ways to deal with a definitive reality and empowering discourse among various mysterious practices.

Theorophical points of view on strict pluralism cross with more extensive investigations into the idea of truth and the chance of an otherworldly solidarity hidden different strict and philosophical viewpoints. Theorophers investigate whether enchanted encounters, with their accentuation on immediate, unmediated experiences with the heavenly, can act as a binding together power that rises above strict limits. Theorophy imagines a worldwide otherworldliness that embraces the wealth of different mysterious customs while recognizing the principal solidarity that interfaces all of humankind.

All in all, the investigation of mysterious encounters inside Theorophy envelops an immense and unpredictable embroidery of thoughts, convictions, and practices. From the old insight of Eastern customs to the pondering bits of knowledge of Western spiritualists, Theorophy winds around together a comprehensive comprehension of reality that embraces both the hypothetical and enchanted components of presence. Theorophers explore the inexpressible idea of magical experiences, wrestling with inquiries of cognizance, time, morals, approval, orientation, language, imagery, and strict pluralism.

Theorophy welcomes people to set out on a groundbreaking excursion that rises above the limits of traditional reasoning, cultivating a more profound association with the secrets of presence. A dynamic and developing field draws in with old insight and contemporary experiences, trying to coordinate the hypothetical and enchanted parts of human cognizance.

In the blend of hypothesis and supernatural quality, Theorophy offers a significant investigation of the idea of the real world and the human mission for importance and reason in the huge span of the universe.

8.1 Examining the role of mystical experiences in Theorophical practices.

Looking at the job of mysterious encounters in Theorophical rehearses uncovers a diverse investigation that digs into the crossing points of reasoning, otherworldliness, and the unspeakable. Theorophy, as a term begat to epitomize the combination of hypothetical and mysterious components, tries to comprehend and explore the domains of presence outside the ability to comprehend of customary language and objective request. In this assessment, we navigate the rich embroidery of Theorophy, wherein otherworldly encounters assume a significant part in molding the comprehension of the real world, cognizance, and the interconnectedness, everything being equal.

At the core of Theorophy lies the acknowledgment that magical encounters are not segregated peculiarities but rather necessary to the texture of human awareness. The investigation of these encounters goes past a simple scholarly activity; it turns into an excursion into the profundities of one's own being and the idea of the universe. Theorophers place that enchanted experiences offer an immediate fear of insights that rise above the constraints of calculated thought, giving bits of knowledge into the central idea of the real world.

In the huge scene of Theorophical thought, Eastern customs have for quite some time been compelling in molding the talk on magical encounters. Hinduism, with its rich embroidery of otherworldly practices, digs into the idea of oneself (Atman) and its definitive solidarity with the heavenly reality (Brahman). The Upanishads, antiquated texts that structure the philosophical groundwork of Hinduism, discuss significant mysterious experiences achieved through reflection and self-request. Theorophy draws from these customs, considering the general rules that rise up out of the supernatural scenes of Eastern idea.

Likewise, Buddhism, with its accentuation on contemplation and care, offers a special point of view on the idea of cognizance and the way to illumination. The magical encounters of Buddhist experts, from the profound conditions of focus (jhana) to the immediate acknowledgment of void (shunyata), add to the Theorophical comprehension of the idea of the psyche and its relationship to extreme reality. Theorophy draws in with Buddhist idea, investigating the ramifications of temporariness, enduring, and non-self as starting points for a Theorophical perspective.

Taoism, established in old Chinese way of thinking, gives one more focal point through which Theorophy ponders supernatural encounters. The Tao Te Ching, a central text in Taoism, discusses the unspeakable Tao as the source and guideline of all presence.

Theorophers dig into the Taoist comprehension of wu-wei (non-activity) and the normal progression of the Tao, thinking about how these ideas line up with the otherworldly encounters of unity and interconnectedness.

In the Western custom, Theorophy connects profoundly with the enchanted components of Christianity, Kabbalah, and Sufism. Christian spiritualists, like Meister Eckhart, John of the Cross, and Julian of Norwich, depict experiences with the heavenly that rise above the limits of traditional strict conventions. Theorophy investigates the groundbreaking idea of these encounters, addressing how they add to a more extensive comprehension of the human relationship with the heavenly.

Kabbalah, the supernatural part of Judaism, offers a rich embroidery of emblematic and thoughtful practices pointed toward accomplishing direct information on God. Theorophy digs into the recondite lessons of Kabbalah, investigating how the mysterious encounters of Jewish spiritualists add to the Theorophical talk on the idea of the heavenly and the design of the real world.

Sufism, the otherworldly custom inside Islam, stresses the immediate experience of God's affection and presence. Sufi spiritualists, like Rumi and Ibn Arabi, portray the excursion of the spirit toward association with the Adored. Theorophy draws in with Sufi lessons, examining the idea of heavenly love and the groundbreaking force of elated conditions of awareness.

The coordination of magical encounters into Theorophy isn't an excusal of reason or scholarly meticulousness. All things considered, it perceives the constraints of a simply rationalistic methodology and embraces the instinctive, experiential elements of human cognizance. Theorophers contend that mysterious encounters give an immediate experience parts of reality that escape the insightful psyche, welcoming people to go past calculated systems and draw in with the secrets of presence at a more profound level.

One of the vital subjects in Theorophical conversations on supernatural encounters is the idea of awareness. Supernatural experiences frequently include a significant change in cognizance, rising above common methods of discernment and comprehension. Theorophy considers whether cognizance itself is an essential part of the real world, existing past the limits of the material mind. The investigation of adjusted conditions of cognizance, from thoughtful retention to hallucinogenic encounters, adds to the Theorophical investigation into the nature and extent of awareness.

The fact that Theorophy basically looks at makes the deterioration of the self image, a common subject in mysterious encounters, another perspective. Spiritualists frequently portray a feeling of egolessness, where the limits among self and other, subject and item, disintegrate.

Theorophers ponder the ramifications of self image disintegration for the comprehension of personality and selfhood, addressing whether the self image is a build that darkens a more profound, more major solidarity with the universe.

Besides, Theorophy draws in with whether or not magical encounters give certified bits of knowledge into the idea of the real world or on the other hand assuming they are results of mental and brain processes. While neuroscience might offer clarifications for specific parts of supernatural encounters, Theorophy accentuates the abstract, individual nature of these experiences, declaring that the full profundity of enchanted bits of knowledge can't be caught by logical investigation alone. Theorophy advocates for an integrative methodology that joins experimental examination with contemplative investigation, perceiving the restrictions of reductionist ways to deal with the investigation of cognizance.

The connection among time and enchanted encounters is one more captivating element of Theorophy. Numerous spiritualists depict an immortal quality to their experiences with the heavenly, where standard ideas of past, present, and future become insignificant. Theorophy considers the ramifications of this immortal aspect, addressing whether it recommends a more profound reality past the transient imperatives of our regular experience.

Theorophical conversations on time and mystery cross with cosmological requests, considering the starting points and extreme destiny of the universe. Some Theorophers suggest that mysterious encounters give looks into an immortal, timeless reality that rises above the limits of the actual universe. Others investigate the possibility that time itself is a build of human insight, and mysterious encounters offer an immediate experience with an immortal, ever-present component of the real world.

While Theorophy draws motivation from antiquated astuteness, it likewise connects with contemporary philosophical and logical viewpoints. The exchange among mystery and quantum material science, for instance, investigates the equals between the otherworldly comprehension of interconnectedness and the quantum idea of non-region. Theorophy embraces the continuous disclosures in the area of neuroscience, perceiving the potential for logical bits of knowledge to supplement and advance the Theorophical investigation of awareness and mysterious encounters.

Theorophy isn't restricted to recondite or conceptual requests; it stretches out its look to the cultural and moral components of supernatural encounters. The job of otherworldliness in the development of strict and profound customs turns into a point of convergence of Theorophical request. Theorophy investigates how magical experiences have formed the moral, moral, and social structures of different social orders since forever ago.

With regards to coordinated religions, Theorophers examine the connection among spiritualists and strict specialists. The pressure between the people who guarantee immediate, unmediated admittance to the heavenly and the individuals who try to control and decipher such encounters is a common subject inside Theorophy. Theorophers dive into the manners by which otherworldly encounters challenge or build up laid out strict teachings, prompting the advancement of strict idea and practice.

Moral contemplations inside Theorophy stretch out past the cultural level to the singular's moral and profound turn of events. Magical encounters are frequently portrayed as impetuses for significant individual change, inciting people to rethink their qualities, needs, and feeling of direction. Theorophy investigates the moral ramifications of such groundbreaking encounters, taking into account how they might impact one's relationship with oneself, others, and the more extensive environmental setting.

Besides, Theorophy ponders the job of otherworldly encounters in the development of temperances like lowliness, empathy, and astuteness. Spiritualists across customs frequently discuss a profound feeling of interconnectedness with all of creation, encouraging an awareness of others' expectations and care for the prosperity of others. Theorophers investigate whether supernatural experiences can act as an establishment for moral rules that rise above social and strict limits, adding to a more comprehensive and merciful worldwide ethic.

Theorophy likewise wrestles with the difficulties and contentions encompassing the translation and approval of otherworldly encounters. Cynics might excuse such encounters as simple mental trips or daydreams, crediting them to mental or neurological elements. Theorophers answer by featuring the multifaceted consistency of otherworldly records and the groundbreaking effect these encounters can have on people and social orders.

In resolving the topic of approval, Theorophy perceives the innately abstract nature of otherworldly encounters. While underscoring the significance of individual legitimacy and genuineness in relating these experiences, Theorophers likewise investigate the job of public and social approval. Theorophical viewpoints on the collective part of otherworldliness consider how shared customs, images, and stories add to the approval and understanding of enchanted encounters inside a specific social or strict setting.

The convergence of orientation and mystery is one more urgent component of Theorophy. Theorophers basically look at authentic and social portrayals of spiritualists, taking into account how orientation elements impact the translation and acknowledgment of supernatural encounters. Theorophy looks to enhance voices that have been underestimated or rejected from customary stories, investigating the commitments of ladies spiritualists and the effect of orientation jobs on the articulation and gathering of mysterious experiences.

Theorophical conversations on orientation and mystery reach out past the double structure, recognizing the variety of orientation personalities and articulations. The interconnection of orientation with different parts of character, like race, class, and sexuality, is a focal concentration inside Theorophy. Theorophers investigate how different social and social elements converge to shape the encounters and translations of magic among people with assorted characters.

In the mission to coordinate otherworldly encounters into Theorophy, the job of language and imagery turns into a focal subject. Enchanted experiences are much of the time depicted as indescribable, past the limit of standard language to catch or convey the experience completely. Theorophers wrestle with the limits of language, perceiving that the actual idea of supernatural encounters difficulties traditional methods of articulation.

Imagery and allegory arise as fundamental apparatuses inside Theorophy to explore the unspeakable idea of magical experiences. Theorophers draw upon strict, social, and prototype images to explain parts of the supernatural experience that oppose clear depiction. The utilization of moral story and legend inside Theorophy conveys the groundbreaking power and significant experiences that rise out of mysterious commitment with a definitive reality.

The connection among supernatural quality and imaginative articulation is likewise investigated inside Theorophy. Specialists, writers, and artists frequently draw motivation from enchanted encounters, utilizing their particular mediums to convey the otherworldly and unutterable components of the real world. Theorophers consider how workmanship fills in as a scaffold between the mysterious and the ordinary, welcoming people to mull over and draw in with the secrets of presence through tasteful encounters.

With regards to Theorophy, the job of imagery stretches out past creative articulation to envelop the translation of strict and fanciful texts. Theorophers participate in hermeneutical requests, investigating how hallowed sacred texts and fantasies encode mysterious experiences and act as guides for those looking to explore the otherworldly way. Theorophy energizes a dynamic and nuanced way to deal with strict texts, perceiving the different layers of importance and the potential for otherworldly insight inside these consecrated stories.

Theorophy likewise wrestles with the subject of strict pluralism and the concurrence of assorted otherworldly practices. Theorophers investigate the shared characteristics and contrasts among different otherworldly ways, trying to observe widespread rules that underlie the assorted articulations of the magical experience. Theorophy advances a feeling of inclusivity, perceiving the legitimacy of numerous ways to deal with a definitive reality and empowering discourse among various magical practices.

Theorophical viewpoints on strict pluralism converge with more extensive investigations into the idea of truth and the chance of an extraordinary solidarity fundamental different strict and philosophical viewpoints. Theorophers investigate whether supernatural encounters, with their accentuation on immediate, unmediated experiences with the heavenly, can act as a binding together power that rises above strict limits. Theorophy imagines a worldwide otherworldliness that embraces the extravagance of different mysterious customs while recognizing the central solidarity that interfaces all of mankind.

All in all, the assessment of the job of mysterious encounters in Theorophical rehearses unfurls as an exhaustive investigation that navigates the domains of reasoning, otherworldliness, and the unspeakable. From the old insight of Eastern practices to the thoughtful bits of knowledge of Western spiritualists, Theorophy winds around together an all encompassing comprehension of reality that embraces both the hypothetical and enchanted elements of presence. Theorophers explore the indescribable idea of mysterious experiences, wrestling with inquiries of cognizance, time, morals, approval, orientation, language, imagery, and strict pluralism.

Theorophy welcomes people to set out on an extraordinary excursion that rises above the limits of traditional reasoning, cultivating a more profound association with the secrets of presence. A dynamic and developing field draws in with old insight and contemporary bits of knowledge, looking to coordinate the hypothetical and supernatural parts of human cognizance. In the combination of hypothesis and magic, Theorophy offers a significant investigation of the idea of the real world and the human journey for importance and reason in the immense spread of the universe.

8.2 Exploring the nature of spiritual revelations and insights.

Investigating the idea of otherworldly disclosures and bits of knowledge uncovers a significant excursion into the profundities of human cognizance, rising above the limits of normal insight. These otherworldly disclosures, frequently portrayed as groundbreaking experiences with the heavenly or extreme reality, comprise a focal topic in different strict, enchanted, and philosophical customs. This investigation looks to explore the rich embroidery of profound encounters, mulling over their suggestions for figuring out the idea of presence, oneself, and the otherworldly.

At its center, the investigation into otherworldly disclosures recognizes the different manners by which people across societies and ages have revealed experiences with a reality past the common. Whether outlined as dreams, disclosures, or mysterious encounters, these experiences share a consistent idea of rising above the impediments of regular mindfulness. Theorophy, a term including the coordination of hypothetical and otherworldly components, gives a system to drawing in with these encounters inside a philosophical setting.

The idea of otherworldly disclosures frequently includes an immediate dread of bits of insight that outperform the grip of normal idea or tangible discernment. In the Eastern customs, the idea of darshan in Hinduism, for instance, means the immediate vision or impression of a god or heavenly reality. Such experiences are viewed as extraordinary, giving significant bits of knowledge into the idea of oneself and its relationship to a definitive reality.

Essentially, in the Western magical practices, profound disclosures are frequently connected with dreams and experiences with the heavenly. The works of spiritualists like Teresa of Avila, Hildegard of Bingen, and St. John of the Cross enlighten the extraordinary idea of their experiences with the heavenly. These disclosures are not simple scholarly experiences but rather envelop a profound, experiential commitment with the secrets of presence.

Theorophy draws in with whether or not otherworldly disclosures offer certifiable bits of knowledge into the idea of the real world or then again on the off chance that they are abstract, socially molded peculiarities. Doubters might see these encounters as results of mental or neurological cycles, crediting them to the human limit with regards to creative mind or modified conditions of cognizance. Theorophy, while recognizing the emotional idea of profound encounters, fights that their extraordinary effect and the consistency of specific subjects across assorted societies warrant serious thought.

The investigation of otherworldly disclosures inside Theorophy dives into the idea of awareness and its ability to rise above the common limits of discernment. Other-worldly encounters frequently include an elevated condition of mindfulness, a feeling of interconnectedness, and a disintegration of the self image. Theorophers examine whether these modified conditions of cognizance give looks into a more profound reality or an extended comprehension of oneself.

The connection between otherworldly disclosures and the idea of inner self disintegration is a common subject inside Theorophy. Spiritualists frequently portray an encounter of rising above the inner self, where the limits among self and other, subject and item, break up. Theorophy questions whether this disintegration of the self image is a brief state or on the other hand on the off chance that it uncovers a key truth about the idea of individual character and its association with a bigger, widespread cognizance.

Besides, Theorophy investigates the fleeting element of profound disclosures. Numerous spiritualists report a feeling of immortality during their experiences with the heavenly, where past, present, and future lose their regular importance. Theorophical conversations on time and otherworldly encounters ponder whether this immortal aspect focuses to a more profound reality past the imperatives of straight time.

Theorophy additionally draws in with whether or not profound disclosures add to the moral and moral advancement of people and social orders. Magical encounters are frequently depicted as impetuses for individual change, driving people to reconsider their qualities and needs.

Theorophy mulls over whether these extraordinary experiences convert into an increased feeling of sympathy, compassion, and moral obligation.

Moral contemplations inside Theorophy stretch out past individual advancement to the cultural level. The job of magic in molding social and strict standards is analyzed, taking into account how otherworldly disclosures impact aggregate qualities and moral systems. Theorophy investigates whether the bits of knowledge acquired through profound encounters can possibly cultivate a more merciful and agreeable worldwide local area.

Theorophers likewise wrestle with the difficulties of deciphering and approving otherworldly disclosures. Wariness frequently emerges in regards to the genuineness of these encounters, with pundits ascribing them to mental or social impacts. Theorophy recognizes the trouble of dispassionately approving emotional encounters however underscores the significance of social, authentic, and mutual settings in understanding the importance and importance ascribed to otherworldly disclosures.

With regards to strict pluralism, Theorophy thinks about the conjunction of different otherworldly customs and their individual cases to truth. The investigation of profound disclosures inside Theorophy doesn't look to lay out a progressive system of strict or otherworldly encounters however empowers a conscious exchange among various practices. Theorophers think about whether otherworldly disclosures, in spite of their social and strict variety, may highlight widespread bits of insight or model examples of human experience.

The convergence of orientation and profound disclosures is one more aspect inside Theorophy. Theorophers fundamentally analyze verifiable and social portrayals of spiritualists, taking into account how orientation elements impact the translation and acknowledgment of profound encounters. Theorophy looks to intensify the voices of ladies spiritualists who might have been minimized or rejected from conventional accounts, investigating the effect of orientation jobs on the articulation and gathering of otherworldly bits of knowledge.

Theorophical conversations on orientation and profound disclosures stretch out past a paired system, recognizing the variety of orientation personalities and articulations. The multifacetedness of orientation with different parts of personality, like race, class, and sexuality, turns into a focal concentration inside Theorophy. Theorophers investigate how different social and social elements meet to shape the encounters and understandings of profound disclosures among people with assorted personalities.

The investigation of otherworldly disclosures inside Theorophy additionally dives into the job of language and imagery in articulating these encounters. Magical experiences are much of the time portrayed as indescribable, past the limit of conventional language to catch or convey the profundity of the experience completely.

Theorophers wrestle with the impediments of language, perceiving that the actual idea of otherworldly disclosures challenges regular methods of articulation.

Imagery and allegory arise as vital devices inside Theorophy to explore the inexpressible idea of profound encounters. Theorophers draw upon strict, social, and original images to verbalize parts of otherworldly disclosures that resist direct portrayal. The utilization of moral story and legend inside Theorophy effectively conveys the extraordinary power and significant experiences that rise up out of otherworldly commitment with a definitive reality.

The connection between profound disclosures and imaginative articulation is likewise investigated inside Theorophy. Specialists, writers, and artists frequently draw motivation from otherworldly encounters, utilizing their individual mediums to convey the extraordinary and inexpressible elements of the real world. Theorophers consider how craftsmanship fills in as an extension between the profound and the regular, welcoming people to examine and draw in with the secrets of presence through tasteful encounters.

With regards to Theorophy, the job of imagery reaches out past creative articulation to incorporate the understanding of strict and fanciful texts. Theorophers take part in hermeneutical requests, investigating how hallowed sacred writings and fantasies encode profound bits of knowledge and act as guides for those trying to explore the otherworldly way. Theorophy supports a dynamic and nuanced way to deal with strict texts, perceiving the various layers of significance and the potential for extraordinary insight inside these holy stories.

Theorophy additionally wrestles with the subject of how profound disclosures converge with the headways in neuroscience and brain research. While profound encounters have frequently been ascribed to divine experiences, Theorophy recognizes the significance of figuring out the brain and mental underpinnings of these encounters. Theorophers take part in a nuanced exchange, trying to coordinate logical experiences with the experiential and otherworldly elements of profound disclosures.

All in all, the investigation of the idea of otherworldly disclosures and experiences inside Theorophy unfurls as a complex excursion that crosses the domains of reasoning, otherworldliness, and the unspeakable. From the different insight of Eastern and Western practices to the pensive bits of knowledge of spiritualists and thinkers, Theorophy winds around together a thorough comprehension of profound encounters. Theorophers explore the complexities of cognizance, time, morals, approval, orientation, language, imagery, and social variety in their thought of profound disclosures.

8.3 Discussing how mysticism enhances the understanding of Theorophical principles.

Examining how enchantment upgrades the comprehension of Theorophical standards unwinds a nuanced investigation at the convergence of reasoning, otherworldliness, and the indescribable. Theorophy, as a discipline that coordinates hypothetical and mysterious components, perceives the characteristic association between scholarly request and direct experiential commitment with extreme reality. This conversation dives into the manners by which enchantment improves and extends Theorophical standards, offering bits of knowledge that rise above the bounds of conventional language and normal talk.

At the center of Theorophical standards lies the acknowledgment that reality stretches out outside the ability to comprehend of experimental perception and calculated understanding. Supernatural quality, with its accentuation on immediate, unmediated experiences with the heavenly or extreme reality, gives a correlative aspect to Theorophy. Theorophers battle that mystery fills in as a wellspring of experiential insight, offering a novel vantage point from which to consider the idea of presence, cognizance, and the interconnectedness, everything being equal.

Eastern customs, with their rich mysterious legacy, assume a critical part in forming Theorophical standards. The Upanishads, fundamental texts in Hinduism, discuss the immediate acknowledgment of oneself (Atman) and its definitive solidarity with the preeminent reality (Brahman). Theorophy draws from these enchanted experiences, pondering the ramifications for the idea of awareness and the self inside the immense region of the universe.

Buddhism, another compelling Eastern custom, adds to Theorophical standards through its investigation of the idea of misery, temporariness, and the way to illumination. The direct experiential methodology of contemplation and care in Buddhism lines up with Theorophy's accentuation on rising above standard methods of discernment to catch key bits of insight about the real world and awareness.

Taoism, established in Chinese way of thinking, acquaints Theorophers with the idea of the Tao as the fundamental standard of all presence. The otherworldly component of Taoism, with its accentuation on immediacy and normal congruity, enhances Theorophical conversations on the interconnectedness of all things and the unfurling of reality as per a grandiose request.

The combination of enchantment into Theorophy likewise includes a basic assessment of Western otherworldly practices. Christian spiritualists, like Meister Eckhart and John of the Cross, portray experiences with the heavenly that go past doctrinal plans. Theorophy draws in with Christian otherworldliness, investigating how these encounters add to a more extensive comprehension of the connection between the singular soul and the extraordinary reality.

Kabbalah, the supernatural part of Judaism, offers representative and pondering practices pointed toward accomplishing direct information on God. Theorophy digs into the elusive lessons of Kabbalah, investigating how the enchanted encounters of Jewish spiritualists illuminate Theorophical standards in regards to the heavenly nature and the design of the real world.

Sufism, the supernatural custom inside Islam, stresses the immediate experience of God's affection and presence. Sufi spiritualists, like Rumi and Ibn Arabi, give significant bits of knowledge into the idea of heavenly solidarity and the spirit's excursion toward association with the Darling. Theorophy draws in with Sufi lessons, considering the groundbreaking force of elated conditions of cognizance and the ramifications for Theorophical standards.

Theorophical standards are additionally enhanced by the assessment of how magic converges with the idea of cognizance. Enchanted encounters frequently include a significant change in cognizance, rising above customary methods of discernment and comprehension. Theorophy ponders whether cognizance itself is a key part of the real world, existing past the limits of the material cerebrum.

The disintegration of the self image, a typical topic in magical encounters, turns into a critical mark of request inside Theorophy. Spiritualists frequently portray a feeling of egolessness, where the limits among self and other, subject and item, break up. Theorophers examine the ramifications of inner self disintegration for the comprehension of personality and selfhood, addressing whether the inner self is a build that clouds a more profound, more major solidarity with the universe.

Theorophy investigates the extraordinary capability of enchanted encounters in molding moral and moral standards. Spiritualists across customs frequently discuss a profound feeling of interconnectedness with all of creation, cultivating an awareness of others' expectations and care for the prosperity of others. Theorophers examine whether magical bits of knowledge can act as an establishment for moral rules that rise above social and strict limits, adding to a more comprehensive and sympathetic worldwide ethic.

With regards to coordinated religions, Theorophy investigates the connection among spiritualists and strict specialists. The strain between the people who guarantee immediate, unmediated admittance to the heavenly and the individuals who look to control and decipher such encounters is a repetitive subject inside Theorophy. Theorophers dig into the manners by which enchanted encounters challenge or support laid out strict regulations, prompting the advancement of strict idea and practice.

Moreover, Theorophy draws in with the extraordinary idea of otherworldly encounters at the singular level. Otherworldly experiences are frequently portrayed as impetuses for significant individual change, inciting people to reexamine their qualities, needs, and feeling of direction.

Theorophy investigates the moral ramifications of such groundbreaking encounters, taking into account how they might impact one's relationship with oneself, others, and the more extensive biological setting.

The investigation of otherworldly encounters inside Theorophy likewise prompts a thought of the idea of time. Numerous spiritualists depict an immortal quality to their experiences with the heavenly, where customary ideas of past, present, and future become immaterial. Theorophy mulls over the ramifications of this immortal aspect, addressing whether it proposes a more profound reality past the transient limitations of our ordinary experience.

Theorophical conversations on time and magic meet with cosmological requests, pondering the starting points and extreme destiny of the universe. Some Theorophers recommend that magical encounters give looks into an immortal, everlasting reality that rises above the limits of the actual universe. Others investigate the possibility that time itself is a build of human insight, and otherworldly encounters offer an immediate experience with an immortal, ever-present element of the real world.

Theorophy perceives the significance of language and imagery in articulating the unspeakable idea of supernatural experiences. Otherworldly encounters are frequently portrayed as rising above normal language, testing the constraints of calculated thought. Theorophers wrestle with the subject of how language, with its innate imperatives, can sufficiently convey the profundity and lavishness of otherworldly bits of knowledge.

Imagery and allegory become fundamental instruments inside Theorophy to explore the inexpressible idea of supernatural experiences. Theorophers draw upon strict, social, and prototype images to express parts of the mysterious experience that resist clear portrayal. The utilization of purposeful anecdote and fantasy inside Theorophy effectively conveys the extraordinary power and significant experiences that rise up out of enchanted commitment with a definitive reality.

Creative articulation likewise turns into a scaffold among mystery and Theorophy. Craftsmen, artists, and performers frequently draw motivation from otherworldly encounters, utilizing their individual mediums to convey the extraordinary and inexpressible components of the real world. Theorophers consider how workmanship fills in as a strong method for correspondence, welcoming people to examine and draw in with the secrets of presence through stylish encounters.

With regards to Theorophy, the job of imagery stretches out past creative articulation to incorporate the understanding of strict and legendary texts. Theorophers participate in hermeneutical requests, investigating how consecrated sacred writings and fantasies encode mysterious experiences and act as guides for those trying to explore the profound way.

Theorophy supports a dynamic and nuanced way to deal with strict texts, perceiving the different layers of importance and the potential for extraordinary insight inside these consecrated stories.

Theorophy, while drawing motivation from old supernatural customs, stays open to the bits of knowledge of contemporary way of thinking and science. The exchange among enchantment and quantum material science, for instance, investigates the equals between the supernatural comprehension of interconnectedness and the quantum idea of non-area. Theorophy embraces the continuous disclosures in the area of neuroscience, perceiving the potential for logical bits of knowledge to supplement and advance the Theorophical investigation of awareness and enchanted encounters.

Theorophy, as a dynamic and developing field, empowers a combination of old insight and current comprehension. The combination of magical encounters into Theorophical standards isn't a dismissal of reason or scholarly meticulousness; all things being equal, it is an acknowledgment of the impediments of a simply rationalistic methodology. Theorophers contend that enchanted encounters furnish an immediate experience with parts of reality that escape the logical psyche, welcoming people to go past reasonable systems and draw in with the secrets of presence at a more profound level.

Chapter 9

The Future of Theorophy

The eventual fate of reasoning is a powerful scene formed by a horde of variables, from innovative progressions to cultural movements. As we explore the intricacies of the 21st hundred years, the job of reasoning turns out to be progressively urgent in giving bits of knowledge and structures to grasping our quickly impacting world. In this investigation representing things to come of reasoning, we dig into key subjects like the reconciliation of arising innovations, the developing idea of human cognizance, the multifacetedness of theory with different disciplines, and the continuous mission for a bringing together hypothesis of everything.

One of the characterizing elements representing things to come of reasoning is its coordination with state of the art innovations. As man-made consciousness (simulated intelligence) keeps on propelling, rationalists are confronted with the test of wrestling with the moral ramifications of canny frameworks. Questions encompassing the ethical organization of artificial intelligence, the idea of awareness in AI, and the possible effect on human character are at the front of philosophical request. The convergence of reasoning and innovation stretches out past morals to investigate the epistemological ramifications of man-made intelligence, inciting us to reevaluate the idea of information and the limits of human perception.

In addition, the approach of computer generated reality (VR) and expanded reality (AR) acquaints new aspects with philosophical talk. Savants are faced with the undertaking of figuring out the ramifications of vivid advanced encounters on discernment, reality, and the development of significance. The obscuring of limits between the virtual and the genuine difficulties conventional philosophical ideas, provoking a reconsideration of primary thoughts like truth, presence, and the idea of reality itself.

As innovation keeps on molding our reality, the fate of reasoning likewise includes a developed commitment with natural way of thinking. The dire difficulties of environmental change, asset consumption, and natural corruption request philosophical systems that can direct mankind towards economical and moral practices. Natural way of thinking turns into a focal field for resolving inquiries of intergenerational equity, the ethical remaining of non-human elements, and the moral obligations of people as stewards of the planet.

As well as wrestling with mechanical and ecological difficulties, the eventual fate of reasoning is complicatedly associated with the advancing idea of human awareness. The interdisciplinary field of neurophilosophy gains noticeable quality as savants team up with neuroscientists to disentangle the secrets of the psyche.

Inquiries regarding the idea of cognizance, the connection among psyche and body, and the chance of counterfeit awareness move reasoning into the front of logical request.

The combination of reasoning and neuroscience leads to moral contemplations in regards to mental upgrade advancements. As we gain the capacity to control and improve mental capabilities, thinkers should explore the moral ramifications of adjusting human instinct. Conversations about the limits of human upgrade, the potential for mental imbalance, and the moral utilization of neurotechnologies become basic to the fate of philosophical talk.

Moreover, the eventual fate of reasoning includes an increased familiarity with social and social settings. The field turns out to be progressively interconnected, recognizing the variety of human encounters and viewpoints. Thinkers draw in with issues of race, orientation, sexuality, and personality, perceiving the interconnectedness of philosophical request with more extensive social battles. The decolonization of reasoning turns into a vital development, testing conventional Eurocentric points of view and encouraging a more comprehensive and various philosophical scene.

In the domain of political way of thinking, what's to come sees a reconsideration of administration structures and the job of the state. As worldwide difficulties require transnational collaboration, savants investigate the potential outcomes of cosmopolitanism and worldwide administration. Inquiries of equity, equity, and basic liberties become the dominant focal point as logicians wrestle with the moral underpinnings of political establishments in an interconnected world.

Besides, the joining of reasoning with arising disciplines like man-made brainpower morals, bioethics, and information morals becomes basic. Thinkers team up with researchers, architects, and policymakers to mindfully foster moral structures that guide innovative progressions. The diversity of reasoning with these fields guarantees that moral contemplations are implanted in the turn of events and arrangement of new advancements, forestalling unseen side-effects and alleviating moral dangers.

The eventual fate of reasoning likewise involves a reconsideration of mystical and epistemological establishments. Logicians investigate the idea of the real world, the constraints of human information, and the chance of a brought together hypothesis that makes sense of the crucial standards of the universe. The mission for a great bringing together hypothesis, frequently sought after inside the domains of power and hypothetical physical science, draws in scholars in exchange with researchers to overcome any issues between experimental perceptions and calculated understanding.

The continuous discourse among reasoning and science turns into a sign representing things to come scholarly scene. As logical disclosures challenge existing philosophical standards, logicians adjust and add to the advancing talk.

The collaboration between these disciplines impels humankind towards a more thorough comprehension of the universe, recognizing the corresponding jobs of experimental examination and reasonable investigation.

In the domain of existential way of thinking, what's in store includes an extended investigation of significance and reason in an undeniably complicated and interconnected world. As people wrestle with inquiries of personality and having a place, existential savants give systems to exploring the existential difficulties of the cutting edge time. The quest for significance in a mechanically intervened presence, the effect of virtual entertainment on self-discernment, and the journey for realness become central places of existential request.

The fate of reasoning reaches out past customary scholarly settings, embracing public way of thinking for the purpose of cultivating discourse and decisive reasoning in the public eye. Rationalists effectively take part openly talk, drawing in with major problems and offering philosophical points of view to a more extensive crowd. The availability of philosophical thoughts through different media channels, including digital recordings, recordings, and online entertainment, democratizes philosophical talk and welcomes assorted voices into the discussion.

Moreover, what's to come sees the proceeded with advancement of moral artificial intelligence and the improvement of man-made intelligence frameworks that can participate in philosophical thinking. The cooperation among rationalists and man-made intelligence frameworks brings up captivating issues about the idea of philosophical ability, the job of instinct in moral direction, and the potential for artificial intelligence to contribute novel experiences to philosophical request. The mix of man-made intelligence into reasoning difficulties conventional thoughts of human excellence and grows the opportunities for cooperative information creation.

The eventual fate of reasoning likewise includes a reexamination of instructive standards, with an emphasis on encouraging philosophical reasoning since the beginning. Philosophical request turns into a fundamental part of training, developing decisive reasoning abilities, moral thinking, and an extended comprehension of the intricacies of the world. The mix of reasoning into instructive educational plans adds to a more educated and intelligent populace, fit for exploring moral quandaries and adding to the thriving of society.

All in all, the fate of reasoning is a dynamic and interdisciplinary scene formed by mechanical headways, ecological difficulties, the developing idea of cognizance, and a pledge to inclusivity. Scholars effectively draw in with arising advancements, wrestle with moral contemplations, and team up across disciplines to resolve the perplexing issues confronting humankind.

The multifacetedness of reasoning with fields like neuroscience, ecological investigations, and man-made consciousness morals guarantees that philosophical request stays significant and receptive to the developing requirements of society. As reasoning keeps on developing, it fills in as a directing light, offering bits of knowledge and structures for exploring the intricacies of the 21st hundred years and then some.

9.1 Speculating on the relevance of Theorophy in the modern world.

Conjecturing on the pertinence of Theorophy in the cutting edge world requires a nuanced assessment of the developing scene in which reasoning works. Theorophy, a term got from the combination of "hypothesis" and "reasoning," embodies the unique idea of philosophical request in the contemporary setting. As we explore the intricacies of the 21st hundred years, the pertinence of Theorophy turns into a point of convergence for understanding and tending to the diverse difficulties that characterize our cutting edge presence.

At the center of Theorophy's importance is its capacity to give reasonable systems to deciphering and figuring out the quickly impacting world. In a time portrayed by mechanical progressions, globalization, and social movements, Theorophy offers a focal point through which people and social orders can draw in with complex issues. The philosophical investigation of key inquiries concerning presence, profound quality, and significance stays a pivotal part of scholarly request, directing us as we continued looking for understanding in a consistently developing worldwide scene.

The significance of Theorophy is especially obvious in its ability to wrestle with the moral ramifications of arising advances. As man-made brainpower, biotechnology, and other logical headways reshape the forms of human experience, Theorophy assumes a crucial part in basically looking at the moral components of these changes. Questions encompassing the ethical organization of man-made intelligence, the ramifications of hereditary designing, and the moral utilization of information are key to Theorophical talk, mirroring the requirement for moral structures that can explore the mind boggling crossing point of innovation and humankind.

Also, Theorophy draws in with the significant changes in human cognizance achieved by mechanical and social movements. The computerized age, described by consistent availability and data over-burden, prompts philosophical investigations into the idea of character, consideration, and the development of the real world. Theorophers investigate the effect of web-based entertainment on self-discernment, the idea of virtual connections, and the difficulties of keeping a lucid identity in a world immersed with computerized improvements.

Theorophy isn't bound to extract consideration however stretches out its importance to tending to squeezing worldwide difficulties, with ecological issues at the very front. The environmental emergency, set apart by environmental change, biodiversity misfortune, and asset exhaustion, requires philosophical reflection on mankind's relationship with the normal world.

Theorophical viewpoints add to natural morals, investigating thoughts of biological obligation, the ethical remaining of non-human substances, and the moral ramifications of human exercises in the world.

In the cutting edge world, the significance of Theorophy is likewise clear in its diversity with different disciplines. The limits between customary scholastic disciplines are progressively permeable, and Theorophy effectively draws in with fields like science, innovation, legislative issues, and financial matters. The coordinated effort among reasoning and these disciplines improves scholarly request, cultivating a more comprehensive comprehension of complicated issues. Theorophy turns into an extension between unique subject matters, working with interdisciplinary exchange and adding to far reaching answers for contemporary difficulties.

Besides, Theorophy adjusts to the developing socio-social scene by embracing variety and inclusivity. As social orders wrestle with issues of civil rights, character, and equity, Theorophy turns into a space for looking at the fundamental philosophical suppositions that shape these discussions. Scholars investigate inquiries of force, honor, and the effect of cultural designs on minimized networks, encouraging a more comprehensive and fair Theorophical talk.

The significance of Theorophy isn't restricted to scholastic circles however reaches out to public talk, where philosophical thoughts add to molding cultural qualities and standards. Public Theorophy turns into a method for democratizing philosophical request, making philosophical ideas open to a more extensive crowd. Theorophers effectively take part out in the open discussions, offering bits of knowledge into major problems, and empowering decisive reasoning past the bounds of scholarly organizations.

In the political domain, Theorophy assumes a significant part in forming philosophies, studying administration designs, and imagining elective socio-political frameworks. The investigation of political way of thinking turns out to be especially relevant as social orders wrestle with inquiries of equity, basic freedoms, and the conveyance of assets. Theorophers add to the advancement of moral structures for administration, pushing for frameworks that focus on human nobility, fairness, and social prosperity.

The pertinence of Theorophy additionally stretches out to the domain of schooling, where philosophical reasoning turns into a fundamental part of developing educated and drew in residents. The mix of Theorophy into instructive educational programs cultivates decisive reasoning abilities, moral thinking, and a more profound comprehension of the intricacies of the world. Theorophical request urges people to address presumptions, examine contentions, and take part in smart reflection, engaging them to explore the moral and scholarly difficulties of the advanced world.

Moreover, Theorophy hypothesizes on the advancing idea of truth and information in the time of data. The advanced upheaval has changed how data is created, spread, and consumed, bringing up basic issues about the idea of truth and the unwavering quality of information. Theorophers investigate ideas like post-truth, deception, and the effect of data innovations on our epistemic works on, adding to a Theorophical comprehension of the developing scene of information.

Theorophy's significance is additionally clear in commitment with existential inquiries emerge notwithstanding mechanical, social, and natural changes. The quest for significance and reason in a quickly developing world turns into a focal subject of

Theorophical request. Theorophers investigate the existential difficulties presented by the cutting edge condition, offering experiences into how people can explore inquiries of character, having a place, and the quest for a significant life.

Also, Theorophy conjectures on the job of sympathy and empathy in tending to worldwide difficulties. In a world set apart by international strains, monetary imbalances, and social treacheries, Theorophers investigate the moral components of human connections and social obligation. Theorophical points of view add to conversations on sympathy as an ethical uprightness, the morals of care, and the opportunities for encouraging a more sympathetic and just society.

As we conjecture on the pertinence of Theorophy in the cutting edge world, recognizing its versatile nature is fundamental. There is no such thing as theorophy in seclusion yet develops in light of the changing shapes of human experience. Theorophers effectively draw in with contemporary issues, presenting philosophical experiences as a powerful influence for the difficulties and chances of the current second.

All in all, the significance of Theorophy in the advanced world is multi-layered and dynamic. It envelops moral contemplations notwithstanding innovative headways, draws in with the changing idea of human awareness, addresses worldwide difficulties like ecological emergencies, and encourages interdisciplinary discourse. Theorophy isn't a remnant of the past yet a no nonsense discipline that keeps on developing, adjust, and add to how we might interpret the intricacies of the 21st 100 years. As we explore the vulnerabilities representing things to come, Theorophy stays a directing power, offering reasonable instruments and philosophical points of view that enhance our aggregate undertaking to figure out the world we possess.

9.2 Discussing potential applications of Theorophical principles in addressing contemporary challenges.

Examining likely uses of Theorophical standards in tending to contemporary difficulties requires a thorough investigation of how philosophical experiences can be tackled to explore the intricacies of the cutting edge world. Theorophy, as a mixture of hypothesis and reasoning, gives a rich establishment to tending to different difficulties spreading over innovation, morals, climate, civil rights, from there, the sky is the limit. In this talk, we dig into the expected uses of Theorophical standards in encouraging moral headways, molding mechanical turns of events, directing natural stewardship, improving civil rights, and advancing all encompassing prosperity.

Moral contemplations pose a potential threat in the domain of mechanical progressions, and Theorophy offers an important focal point through which to explore the moral landscape of the computerized age. The standards of Theorophy can be applied to shape the turn of events and arrangement of arising innovations, like man-made reasoning (artificial intelligence), biotechnology, and information investigation. Theorophical request energizes a smart assessment of the moral ramifications encompassing these innovations, guaranteeing that they line up with human qualities, safeguard security, and try not to sustain unsafe predispositions.

Theorophical standards add to the foundation of moral systems for computer based intelligence, resolving issues of straightforwardness, responsibility, and reasonableness. The investigation of moral simulated intelligence includes contemplations of the ethical organization of astute frameworks, the effect on human personality, and the cultural ramifications of far and wide artificial intelligence reception. By applying Theorophical experiences, we can cultivate an innovation driven scene that focuses on human prosperity, civil rights, and the capable utilization of cutting edge calculations.

Moreover, Theorophy assumes an essential part in directing moral contemplations in the domain of biotechnology. As hereditary designing and other biotechnological progressions become progressively pervasive, Theorophical standards offer an establishment for resolving inquiries concerning the ethical limits of human improvement, the moral utilization of hereditary data, and the ramifications of controlling the structure blocks of life. Theorophical talk gives a space to considering the moral consequences of biotechnological intercessions, guaranteeing that they line up with upsides of equity, value, and regard for human nobility.

In the computerized age, the assortment and examination of huge measures of information raise critical moral worries, going from issues of protection to the expected abuse of individual data. Theorophical standards add to the improvement of information morals, stressing the capable and impartial utilization of information in different spaces, including medical services, money, and administration. Theorophical investigation into information morals thinks about inquiries of assent, straightforwardness, and the fair dissemination of the advantages and dangers related with information driven advancements.

The expected utilizations of Theorophical standards reach out past innovation to address squeezing ecological difficulties. Ecological way of thinking, a part of Theorophy, investigates humankind's relationship with the normal world and gives a system to moral contemplations in natural stewardship. Theorophical standards can direct endeavors to address environmental change, biodiversity misfortune, and asset exhaustion by advancing economical practices, capable asset the board, and a more profound comprehension of our moral obligations to the planet.

Theorophical points of view add to ecological morals, accentuating the interconnectedness of every living being and the inherent worth of nature. The thought of non-human elements as moral subjects, instead of simple assets, lines up with Theorophical rules that promoter for an all encompassing and comprehensive moral structure. Theorophy empowers a shift from anthropocentrism to ecocentrism, perceiving the inborn worth of biological systems and the significance of safeguarding biodiversity for the prosperity of the planet.

Also, Theorophical standards are instrumental in molding moral mentalities towards utilization and way of life decisions. The idea of supportable Theorophy urges people and social orders to embrace rehearses that limit natural effect, decrease squander, and advance environmental agreement. Theorophical experiences add to the improvement of a maintainable ethic that perceives the interconnectedness of biological, social, and monetary frameworks, encouraging an all encompassing way to deal with tending to ecological difficulties.

Chasing after civil rights, Theorophical standards offer important devices for investigating existing power structures, supporting for balance, and tending to fundamental shameful acts. The multifacetedness intrinsic in Theorophy considers a nuanced assessment of the complicated transaction of variables like race, orientation, class, and sexuality in molding individual encounters and cultural designs. Theorophical viewpoints add to the destroying of abusive frameworks by testing instilled predispositions, advancing inclusivity, and upholding for the privileges and pride of underestimated networks.

Theorophy illuminates civil rights developments by giving calculated systems to understanding and resolving issues of imbalance and segregation. The use of Theorophical standards to civil rights includes a basic assessment of cultural standards, social practices, and institutional designs that sustain bad form. Theorophers take part in exchange with activists, policymakers, and networks to add to the improvement of moral arrangements that advance value and inclusivity.

Besides, Theorophy assumes a critical part in forming political way of thinking and administration models that focus on human prosperity and civil rights. The investigation of political Theorophy includes addressing existing political frameworks, supporting for participatory majority rules system, and imagining elective models that focus on the benefit of

everyone. Theorophical standards add to the advancement of moral administration structures that address issues of force, responsibility, and the appropriation of assets in the public eye.

Chasing comprehensive prosperity, Theorophical standards add to philosophical points of view on existential inquiries, psychological well-being, and the quest for a significant life. Theorophy draws in with the difficulties presented by the cutting edge condition, like the effect of innovation on human cognizance, the disintegration of customary conviction frameworks, and the mission for reason in a quickly impacting world. Theorophical experiences add to the investigation of existential inquiries, offering direction on how people can explore the intricacies of present day presence.

Theorophy meets with positive brain research, a field that spotlights on human thriving and prosperity. The utilization of Theorophical standards to positive brain science includes investigating the moral components of joy, the quest for importance, and the development of excellencies that add to individual and aggregate prosperity. Theorophers participate in discourse with analysts, specialists, and teachers to contribute philosophical viewpoints that enhance the comprehension of prosperity and versatility notwithstanding contemporary difficulties.

Also, Theorophical standards illuminate moral contemplations in medical care, bioethics, and the improvement of clinical advances. Theorophy adds to conversations on the ethical ramifications of clinical intercessions, the moral treatment of patients, and the fair dispersion of medical care assets. Theorophical viewpoints guide the improvement of moral structures that focus on human pride, independence, and the mindful utilization of clinical headways chasing further developed wellbeing results.

All in all, talking about the expected uses of Theorophical standards in tending to contemporary difficulties highlights the dynamic and versatile nature of reasoning in the cutting edge world. From molding moral contemplations in innovation and natural stewardship to adding to civil rights developments and directing all encompassing prosperity, Theorophy fills in as a flexible and fundamental instrument for exploring the intricacies of the 21st 100 years. By applying Theorophical bits of knowledge, we can encourage a more moral, comprehensive, and practical world that mirrors the upsides of equity, sympathy, and the quest for a significant life.

9.3 Encouraging readers to explore Theorophy as a source of wisdom and guidance.

Empowering perusers to investigate Theorophy as a wellspring of shrewdness and direction includes featuring the wealth and pertinence of Theorophical standards in exploring the intricacies of life. The expression "Theorophy," a combination of hypothesis and reasoning, epitomizes a unique way to deal with philosophical request that goes past dynamic thought and effectively draws in with the difficulties and questions that characterize the human experience.

In this talk, we dig into the motivations behind why perusers ought to investigate Theorophy, looking at its ability to give astuteness, guide moral navigation, encourage self-improvement, and proposition bits of knowledge into the crucial parts of presence.

At the core of empowering perusers to investigate Theorophy is the acknowledgment of reasoning as a wellspring of insight that rises above the limits of time and social settings. Theorophical standards, drawing from assorted philosophical customs, offer immortal bits of knowledge into the idea of the real world, human cognizance, and the moral components of presence. By diving into Theorophy, perusers get to a repository of gathered human insight that traverses hundreds of years, giving an establishment to smart reflection and a more profound comprehension of the intricacies of life.

Theorophy fills in as an aide for moral dynamic notwithstanding contemporary difficulties. In a world set apart by mechanical progressions, natural emergencies, and civil rights issues, Theorophical standards offer an ethical compass to explore the moral territory. By investigating Theorophy, perusers draw in with philosophical viewpoints on equity, sympathy, and the interconnectedness of every single living being. Theorophy turns into a wellspring of moral direction, empowering perusers to basically look at their qualities, pursue informed decisions, and add to the formation of an all the more and merciful society.

Moreover, Theorophy welcomes perusers to set out on an excursion of self-awareness and self-disclosure. The investigation of basic inquiries concerning presence, importance, and reason turns into a groundbreaking cycle that encourages scholarly development and mindfulness. Theorophical request urges perusers to consider their convictions, values, and presumptions, provoking a more profound comprehension of themselves and their position on the planet. By drawing in with Theorophy, perusers leave on a philosophical investigation that isn't just mentally animating yet additionally improves their own and existential aspects.

Theorophy offers a comprehensive way to deal with understanding the human experience, incorporating scholarly pursuits as well as close to home, moral, and profound aspects. Perusers investigating Theorophy have the chance to draw in with philosophical points of view on adoration, excellence, bliss, and the quest for significance. Theorophical standards add to a more complete comprehension of leading a satisfying and significant life, giving perusers bits of knowledge that go past the outer layer of regular encounters.

Besides, uplifting perusers to investigate Theorophy includes featuring its part in cultivating a feeling of interconnectedness with the more extensive human local area and the normal world. Theorophy stresses the interconnectedness of every living being and the moral obligations that emerge from this interconnectedness. Perusers, through the investigation of Theorophy, can develop a feeling of sympathy, empathy, and biological mindfulness, perceiving the effect of their decisions on the prosperity of others and the planet.

Theorophy isn't restricted to the domain of unique hypothesis; it effectively draws in with contemporary issues and adds to continuous discussions about the fate of humankind. By investigating Theorophy, perusers become members in a discourse that rises above disciplinary limits and addresses the squeezing difficulties of the 21st 100 years. Theorophical standards converge with fields like innovation morals, natural way of thinking, and civil rights, making Theorophy a pertinent and dynamic wellspring of direction for those looking to comprehend and answer the intricacies of the cutting edge world.

Empowering perusers to investigate Theorophy likewise includes scattering the misinterpretation that way of thinking is an elusive or distant pursuit. Theorophy, as a comprehensive and dynamic way to deal with theory, invites perusers from different foundations and encounters. Theorophical request isn't restricted to scholastic circles; it reaches out to public talk, making philosophical experiences open to a more extensive crowd through different media channels, including books, webcasts, and online stages. Perusers are urged to move toward Theorophy as a device for regular living, a wellspring of motivation, and an aide for exploring the moral and existential difficulties they experience.

Theorophy fills in as an extension between various philosophical customs, welcoming perusers to investigate the lavishness and variety of human idea. By drawing in with Theorophy, perusers gain openness to Eastern and Western methods of reasoning, old and contemporary points of view, and a variety of social and philosophical customs. This inclusivity expands the extent of philosophical investigation, permitting perusers to draw motivation from various sources and encouraging a more nuanced and far reaching comprehension of the human experience.

Moreover, Theorophy urges perusers to embrace vulnerability and draw in with the intrinsic intricacy of philosophical inquiries. In a world portrayed by quick change and vulnerability, Theorophy gives a space to investigating vagueness, wrestling with conundrums, and recognizing the limits of human information. Perusers are welcome to move toward Theorophy with a receptive outlook, perceiving that philosophical request isn't tied in with finding conclusive responses yet about taking part in a non-stop course of addressing, reflection, and investigation.

The investigation of Theorophy is an encouragement to develop decisive reasoning abilities and scholarly interest. Theorophical request urges perusers to address presumptions, challenge assumptions, and participate in smart reflection. By effectively taking part during the time spent philosophical request, perusers foster the ability to investigate complex issues, think about various viewpoints, and go with informed choices. Theorophy turns into an impetus for scholarly strengthening, cultivating an outlook that values interest, basic request, and a long lasting obligation to learning.

Empowering perusers to investigate Theorophy likewise includes perceiving its part in encouraging a feeling of marvel and wonderment even with the secrets of presence.

Theorophical standards brief perusers to examine the idea of the real world, the constraints of human comprehension, and the significant inquiries that have captivated masterminds from the beginning of time. Theorophy welcomes perusers to develop a feeling of scholarly lowliness, recognizing the immeasurability of the obscure and the vast opportunities for investigation and revelation.

All in all, uplifting perusers to investigate Theorophy as a wellspring of shrewdness and direction is an encouragement to leave on a groundbreaking excursion of scholarly investigation, moral reflection, and self-awareness. Theorophy, with its dynamic and comprehensive way to deal with reasoning, offers perusers a diverse focal point through which to draw in with the intricacies of the human experience. By diving into Theorophy, perusers have the valuable chance to draw upon immortal experiences, explore contemporary difficulties, and develop a more profound comprehension of themselves and the world they possess. Theorophy coaxes as a wellspring of motivation, an aide for moral direction, and a friend in the continuous mission for shrewdness and importance.

In the unpredictable embroidery of human life, the quest for shrewdness and direction has been an immortal undertaking, woven into the actual texture of our shared perspective. All through the records of history, people and civic establishments have looked for wellsprings of understanding and edification to explore the intricacies of life. This mission for shrewdness rises above social, strict, and philosophical limits, joining humankind in a common excursion towards figuring out the significant secrets that encompass us.

At the core of this mission lies the acknowledgment that life is a dynamic and eccentric excursion, set apart by snapshots of bliss and distress, achievement and disappointment, love and misfortune. Even with such vulnerabilities, the human soul longs for a compass, a directing light that can enlighten the way forward. This natural craving for direction has appeared in different structures, going from antiquated sacred texts and philosophical compositions to oral customs went down through ages.

One of the getting through wellsprings of intelligence and direction is religion. Across the globe, different religions have arisen, each offering an extraordinary viewpoint on the idea of presence and the reason for life. These strict practices frequently give moral codes, moral standards, and accounts that proposition comfort and heading to their adherents. Whether it's the lessons of Buddha, the insight of the Quran, the experiences of the Bhagavad Gita, or the moral rules of Confucianism, strict texts act as reference points of illumination for millions.

Strict pioneers, venerated as courses of heavenly insight, assume a crucial part in deciphering and scattering the lessons of their separate religions. From prophets and saviors to holy people and masters, these otherworldly aides are viewed as delegates between the heavenly and the natural domain. Their words and activities are examined for pieces of ageless insight, and their lives frequently become stories that enlighten the way to otherworldly edification.

However, past the domain of coordinated religion, common ways of thinking and moral structures likewise offer significant bits of knowledge into the human condition. Logicians, masterminds, and researchers have wrestled with central inquiries regarding ethical quality, equity, and the idea of the real world. The works of Aristotle, Plato, Immanuel Kant, Friedrich Nietzsche, and others have molded the scholarly scene, giving systems to moral navigation and reflections on the significance of life.

In the huge woven artwork of human idea, writing arises as one more rich wellspring of shrewdness and direction. Through the ages, writers, authors, and narrators have woven accounts that catch the intricacies of the human experience. Writing has the ability to summon sympathy, incite thoughtfulness, and proposition significant bits of knowledge into the complexities of connections, society, and oneself. From the immortal works of Shakespeare to the existential investigations of Dostoevsky, writing fills in as a mirror mirroring the kaleidoscope of human feelings and difficulties.

Besides, the oral custom, an old type of narrating went down through ages, has been a fundamental wellspring of shrewdness for some social orders. Older folks, griots, and narrators play had a pivotal impact in saving social legacy and sending significant life examples starting with one age then onto the next. Through fantasies, legends, and folktales, networks have granted moral lessons, social qualities, and viable direction for exploring the difficulties of life.

In the cutting edge period, the appearance of innovation has changed the scene of astuteness and direction. The web, with its huge storehouse of data, has democratized admittance to information. Online stages, webcasts, and web-based entertainment interface people with different viewpoints, empowering the trading of thoughts and bits of knowledge on a worldwide scale. Be that as it may, this computerized age likewise presents difficulties, as the wealth of data can be overpowering, and knowing the true from the deceptive turns into a basic expertise.

In the midst of this rich embroidery of sources, the normal world itself arises as a significant educator. Noticing the patterns of nature, the interconnectedness of bio-logical systems, and the strength of living things, one can gather important examples about versatility, equilibrium, and amicability. Native societies, profoundly receptive to the rhythms of the Earth, frequently find otherworldly direction in the normal world, seeing it as a consecrated embroidery that offers shrewdness to the people who tune in.

Human connections, as well, are a cauldron of intelligence and direction. Communications with family, companions, coaches, and even enemies give valuable open doors to development, self-revelation, and the development of temperances like sympathy, compassion, and flexibility. The common encounters of adoration, struggle, and cooperation become a material on which people paint the story of their lives, learning important examples about the human condition en route.

Chasing intelligence and direction, the thoughtful excursion of self-revelation holds a focal spot. The old order "Know thyself" reverberations through the halls of time, welcoming people to leave on an inward mission to grasp their own inspirations, fears, and desires. Practices like contemplation, care, and scrutinizing reflection offer devices for diving into the profundities of cognizance, revealing the layers of oneself, and accomplishing a more clear vision of one's motivation throughout everyday life.

Across societies, the prime example of the savvy senior or sage encapsulates the zenith of a daily existence devoted to the quest for shrewdness. Whether represented as the yogi in the Himalayas, the Local American medication man, or the Eastern savant in profound thought, these figures represent the refined insight that accompanies age, insight, and a significant association with the secrets of presence. Their advice is frequently looked for by those exploring the wild waters of life, and their words convey the heaviness of collected bits of knowledge.

In the domain of science, the mission for astuteness appears as exact request and the steady quest for information. Logical revelations disentangle the secrets of the universe, giving an objective structure to understanding the regulations overseeing the universe. The logical strategy, with its accentuation on proof, perception, and trial and error, has introduced a period of uncommon innovative headways, forming the manner in which humankind collaborates with the world.

However, the logical viewpoint, while significant in disentangling the mechanics of the actual universe, wrestles with innate impediments with regards to resolving existential inquiries concerning the importance and motivation behind life. The reductionist methodology of science, zeroing in on the quantifiable and discernible, may not completely catch the complexities of human awareness, profound quality, and the emotional components of involvement.

In the embroidery of shrewdness and direction, human expressions contribute a one of a kind string, winding around together the domains of creative mind, feeling, and style. Music, visual expressions, dance, and other inventive articulations act as vehicles for investigating the unutterable parts of the human experience. The craftsman, much the same as a soothsayer, distils widespread insights into a medium that rises above language, welcoming people to interface with the grand and the otherworldly.

Moral and moral systems, whether established in strict lessons, philosophical standards, or social qualities, give people rules for exploring the ethical intricacies of life. Ideas of good and bad, equity and decency, sympathy and compassion, structure the moral embroidery that shapes human way of behaving and cultural standards. These ethical compasses act as guides in snapshots of direction, encouraging people to think about the outcomes of their activities on themselves as well as other people.

As social orders develop and confront new difficulties, the aggregate insight of networks turns into an important asset. Native information, went down through ages, frequently holds supportable and amicable practices that resound with the rhythms of the normal world. Conventional recuperating techniques, environmental stewardship, and public customs mirror an interconnected perspective that perceives the reliance of every single living being.

The quest for shrewdness and direction isn't restricted to the individual or cultural level; it reaches out to the worldwide stage. In a period set apart by interconnectedness and relationship, the difficulties confronting humankind — from environmental change to international strains — require cooperative and shrewdness driven arrangements. Global collaboration, educated by a profound comprehension regarding the common predetermination of the human family, becomes basic in exploring the intricacies of the contemporary world.

In the stupendous embroidery of presence, the mission for shrewdness and direction is a consistently unfurling venture. It is an excursion set apart by different scenes, where the pioneer experiences the holy and the everyday, the significant and the normal. The wellsprings of shrewdness are basically as shifted as the human experience itself, mirroring the variety of ways that people and social orders navigate in their quest for understanding.

At its center, the quest for shrewdness is a profoundly human undertaking, driven by an essential longing to figure out the secrets of presence. Whether through strict confidence, philosophical request, logical investigation, creative articulation, or the basic yet significant illustrations of regular day to day existence, people end up attracted to sources that offer lucidity, reason, and an internal compass on their life's process.

Amidst this journey, the idea of an all inclusive insight — an immortal and extraordinary wellspring of direction — arises as a tempting chance. This model, present in different social and strict stories, recommends the presence of a vast insight that pervades the texture of the real world. Whether conceptualized as the Tao, the Logos, or the Grandiose Psyche, this widespread insight epitomizes the possibility that there is a fundamental request to the universe, and by lining up with it, people can track down concordance and significance in their lives.

As the searcher of insight crosses the scenes of presence, they experience difficulties that test their flexibility, situations that request moral acumen, and snapshots of wonderment that welcome consideration. In confronting these encounters, the individual turns into an explorer on an otherworldly excursion, an excursion that rises above the limits of existence, enveloping the total of the human experience.

The incredible sages and shrewdness customs across societies frequently discuss an extraordinary perspective to the quest for insight. It isn't only the collection of information or the adherence to a bunch of standards; rather, it is an excursion of inward speculative chemistry, a course of self-change that unfurls as the searcher draws in with the insight they experience.

This change isn't an objective however a persistent unfurling, a dance between the known and the obscure, the limited and the endless.

In the domain of strict and otherworldly customs, this extraordinary excursion is much of the time outlined as a way of edification, arousing, or freedom. Whether it's the Buddhist idea of Nirvana, the Hindu quest for Moksha, or the Christian thought of salvation, the pith is a significant change in cognizance — an enlivening to a reality that rises above the constraints of the egoic self. The searcher, in this specific situation, moves past the domain of simple scholarly comprehension into an immediate, experiential knowing about the heavenly or a definitive reality.

In philosophical customs, the groundbreaking part of shrewdness is reflected in the idea of intelligence as an uprightness. Insight isn't simply the ownership of information yet the exemplification of ethical characteristics like judiciousness, boldness, restraint, and equity. The savvy individual, as per old style philosophical idea, is one who figures out the intricacies of life as well as encapsulates moral standards in their activities, adding to the prosperity of themselves and society.

The groundbreaking excursion of shrewdness is likewise obvious in artistic expression. Imaginative articulation turns into a medium through which the craftsman draws in with the profundities of their own mind, investigates the subtleties of human inclination, and conveys widespread bits of insight. The course of creation itself is a groundbreaking demonstration, an excursion of self-disclosure and disclosure that makes a permanent imprint on both the craftsman and the crowd.

With regards to science, the quest for shrewdness appears as a constant refinement of understanding. Logical hypotheses, based upon perception and trial and error, go through a course of development as new proof arises. The logical undertaking is definitely not a static collection of realities however a unique investigation of the secrets of the universe, welcoming researchers to constantly address, refine, and extend how they might interpret the universe.

The groundbreaking idea of astuteness is maybe generally substantial in the cauldron of human connections. Connections with others become a mirror reflecting parts of the self that might stay concealed in isolation. Connections, whether familial, heartfelt, or dispassionate, offer open doors for development, sympathy, and the development of ideals like empathy and pardoning. The savvy individual, through their connections, adds to the prosperity of people around them and cultivates a feeling of interconnectedness.

The idea of a general insight proposes that, past the variety of social, strict, and philosophical articulations, there might be a consistent idea that joins the different ways to shrewdness. This general insight, assuming it exists, rises above the constraints of language and doctrine, offering a binding together vision that resounds with the most profound desires of the human soul.

It welcomes people to look past the outer layer of contrasts and perceive the common mission for importance and understanding that ties humankind together.

However, the actual idea of shrewdness opposes simple definition or embodiment. It evades the grip of unbending classes and fixed principles, welcoming people to move toward it with lowliness and transparency. Shrewdness, in its embodiment, is alive, dynamic, and receptive to the always changing scene of human experience. It's anything but a static store of answers however a living presence that entices people to participate in a ceaseless discourse with the secrets of presence.

The quest for shrewdness, then, at that point, is certainly not a lone undertaking however a public and intergenerational undertaking. It includes an exchange between the insight of the past, the bits of knowledge of the present, and the goals representing things to come. The insight customs gave over through ages, whether as sacred texts, philosophical compositions, or oral lessons, act as guideposts for those exploring the intricacies of contemporary life.

In this amazing embroidered artwork of shrewdness, the job of training becomes fundamental. Instructive establishments, from the earliest phases of figuring out how to the most noteworthy echelons of the scholarly community, assume an essential part in forming the points of view and upsides of people. All encompassing training, one that coordinates scholarly, moral, and profound aspects, turns into a vehicle for supporting the seeds of shrewdness in the hearts and brains of the more youthful age.

The subject of how to develop shrewdness with regards to schooling is a mind boggling and diverse one. It includes the transmission of information and abilities as well as the encouraging of decisive reasoning, moral insight, and a feeling of direction. Teachers, as stewards of the learning venture, bear a profound obligation in directing understudies toward a more profound comprehension of themselves, their position on the planet, and the moral ramifications of their activities.

Besides, the coordination of different points of view and perspectives in training becomes significant in a globalized and interconnected world. Openness to the lavishness of social, strict, and philosophical customs encourages an appreciation for the embroidered artwork of human variety. It urges people to move toward the mission for intelligence with an open heart and an eagerness to gain from the horde articulations of human understanding.

The thought of a wellspring of insight and direction infers a relationship — a unique transaction between the searcher and the looked for. In strict practices, this relationship is in many cases outlined as a pledge between the individual and the heavenly. The searcher, through petition, reflection, or custom, looks for fellowship with an otherworldly reality, giving up their egoic self in a journey for divine direction and brightening.

In philosophical customs, the relationship with shrewdness assumes the personality of a persuasive commitment. The logician participates in a nonstop exchange with the extraordinary inquiries of presence, testing suspicions, investigating prospects, and refining their figuring out through the pot of scholarly request.

9 788819 679550 4